GUIDE FOR MAKING ACUTE RISK DECISIONS

GUIDE FOR MAKING ACUTE
RISK DECISIONS

CENTER FOR CHEMICAL PROCESS SAFETY
of the
AMERICAN INSTITUTE OF CHEMICAL ENGINEERS
New York, NY

WILEY

This edition first published 2020
© 2020 the American Institute of Chemical Engineers

A Joint Publication of the American Institute of Chemical Engineers and John Wiley & Sons, Inc.

Registered Office
John Wiley & Sons, Inc., 111 River Street, Hoboken, NJ 07030, USA

Editorial Office
111 River Street, Hoboken, NJ 07030, USA

For details of our global editorial offices, customer services, and more information about Wiley products visit us at www.wiley.com.

Wiley also publishes its books in a variety of electronic formats and by print-on-demand. Some content that appears in standard print versions of this book may not be available in other formats.

Library of Congress Cataloging-in-Publication Data is available.
Hardback ISBN: 9781118930212

Cover Design: Wiley
Cover Images: Silhouette, oil refinery © manyx31/iStock.com; Stainless steel © Creativ Studio Heinemann/Getty Images; Dow Chemical Operations, Stade, Germany/Courtesy of The Dow Chemical Company

Printed in the United States of America

V10014650_101119

CONTENTS

LIST OF TABLES

LIST OF FIGURES

ACRONYMS AND ABBREVIATIONS

ACC	American Chemistry Council
ACGIH	American Conference of Governmental Industrial Hygienists
AEGL	Ambient Air Exposure Guidelines
AIChE	American Institute of Chemical Engineers
AIT	Auto Ignition Temperature
API	American Petroleum Institute
ASME	American Society of Mechanical Engineers
BLEVE	Boiling Liquid Expanding Vapor Explosion
BMS	Burner Management System
CCA	Cause Consequence Analysis
CEI	Chemical Exposure Index (Dow Chemical)
CFR	Code of Federal Registry
CMA	Chemical Manufacturers Association
CSB	US Chemical Safety and Hazard Investigation Board
CCPS	Center for Chemical Process Safety
CCR	Continuous Catalyst Regeneration
COO	Conduct of Operations
CPI	Chemical Process Industries
DCU	Delayed Coker Unit
DDT	Deflagration to Detonation Transition
DIERS	Design Institute for Emergency Relief Systems
ERS	Emergency Relief System
ERPG	Emergency Response Planning Guidelines
EPA	US Environmental Protection Agency
FCCU	Fluidized Catalytic Cracking Unit

F&EI	Fire and Explosion Index (Dow Chemical)
FMEA	Failure Modes and Effect Analysis
HAZMAT	Hazardous Materials
HAZOP	Hazard and Operability Study
HIRA	Hazard Identification and Risk Analysis
HTHA	High Temperature Hydrogen Attack
HRA	Human Reliability Analysis
HSE	Health & Safety Executive (UK)
I&E	Instrument and Electrical
IDLH	Immediately Dangerous to Life and Health
ISD	Inherently Safer Design
IPL	Independent Protection Layer
ISO	International Organization for Standardization
ISOM	Isomerization Unit
ITPM	Inspection Testing and Preventive Maintenance
LFL	Lower Flammable Limit
LNG	Liquefied Natural Gas
LOPA	Layer of Protection Analysis
LOTO	Lock Out Tag Out
LPG	Liquefied Petroleum Gas
MAIT	Maximum Auto Ignition Temperature
MAWP	Maximum Allowable Working Pressure
MCC	Motor Control Center
MEC	Minimum Explosible Concentration
MIE	Minimum Ignition Energy
MOC	Management of Change
MOOC	Management of Organizational Change
MSDS	Material Safety Data Sheet
NASA	National Aeronautics and Space Administration

NDT	Non- Destructive Testing
NFPA	National Fire Protection Association
NPV	Net Present Value
OCM	Organizational Change Management
OIMS	Operational Integrity Management System (ExxonMobil)
OSHA	US Occupational Safety and Health Administration
PHA	Process Hazard Analysis
PLC	Programmable Logic Controllers
PRA	Probabilistic Risk Assessment
PRD	Pressure Relief Device
PRV	Pressure Relief Valve
PSB	Process Safety Beacon
PSE	Process Safety Event
PSI	Process Safety Information
PSI	Process Safety Incident
PSM	Process Safety Management
PSO	Process Safety Officer
PSSR	Pre-Startup Safety Review
QRA	Quantitative Risk Analysis
RBPS	Risk Based Process Safety
RAGAGEP	Recognized and Generally Accepted Good Engineering Practice
RMP	Risk Management Plan
SACHE	Safety and Chemical Engineering Education
SCAI	Safety Controls Alarms and Interlocks
SHE	Safety, Health and Environmental
SHIB	Safety Hazard Information Bulletin
SIS	Safety Instrumented Systems
SME	Subject Matter Expert

TQ   Threshold Quantity

UFL   Upper Flammable Limit

UK   United Kingdom

US   United States

UST   Underground Storage Tank

GLOSSARY

Acute Toxicity
The adverse (acute) effects resulting from a single dose or exposure to a substance. Importance: Ordinarily used to denote effects in experimental animals.

Asset integrity
A PSM program element involving work activities that help ensure that equipment is properly designed, installed in accordance with specifications, and remains fit for purpose over its life cycle. Also see asset integrity and reliability.

Atmospheric Storage Tank
A storage tank designed to operate at any pressure between ambient pressure and 0.5 psig (3.45kPa gage).

Boiling-Liquid-Expanding-Vapor Explosion (BLEVE)
A type of rapid phase transition in which a liquid contained above its atmospheric boiling point is rapidly depressurized, causing a nearly instantaneous transition from liquid to vapor with a corresponding energy release. A BLEVE of flammable material is often accompanied by a large aerosol fireball, since an external fire impinging on the vapor space of a pressure vessel is a common cause. However, it is not necessary for the liquid to be flammable to have a BLEVE occur.

Bow Tie Diagram
A diagram for visualizing the types of preventive and mitigative barriers which can be used to manage risk. These barriers are drawn with the threats on the left, the unwanted event at the center, and the consequences on the right, representing the flow of the hazardous materials or energies through its barriers to its destination. The hazards or threats can be proactively addressed on the left with specific barriers (safeguards, layers of protection) to help prevent a hazardous event from occurring; barriers reacting to the event to help reduce the event's consequences are shown on the right.

Checklist Analysis
A hazard evaluation procedure using one or more pre-prepared lists of process safety considerations to prompt team discussions of whether the existing safeguards are adequate.

Combustible Dust	Any finely divided solid material that is 420 microns or smaller in diameter (material passing through a U.S. No. 40 standard sieve) and presents a fire or explosion hazard when dispersed and ignited in air or other gaseous oxidizer.
Conduct of Operations (COO)	The embodiment of an organization's values and principles in management systems that are developed, implemented, and maintained to (1) structure operational tasks in a manner consistent with the organization's risk tolerance, (2) ensure that every task is performed deliberately and correctly, and (3) minimize variations in performance.
Consequence	The undesirable result of a loss event, usually measured in health and safety effects, environmental impacts, loss of property, and business interruption costs.
Consequence Analysis	The analysis of the expected effects of incident outcome cases, independent of frequency or probability.
Dispersion Models	Mathematical models that characterize the transport of toxic/flammable materials released to the air and/or the water.
Domino Effects	The triggering of secondary events, such as toxic releases, by a primary event, such as an explosion, such that the result is an increase in consequences or area of an effect zone. Generally only considered when a significant escalation of the original incident results.
Emergency Response Planning Guidelines	A system of guidelines for airborne concentrations of toxic materials prepared by the AIHA. For example, ERPG-2 is the maximum airborne concentration below which it is believed nearly all individuals could be exposed for up to one hour without experiencing or developing irreversible or other serious health effects or symptoms that could impair an individual's ability to take protective action.
Event Tree Analysis	A method used for modeling the propagation of an initiating event through the sequence of possible incident outcomes. The event is represented graphically by a tree with branches from the initiating cause through the success or failure of independent protection layers.

Explosion	A release of energy that causes a pressure discontinuity or blast wave.
Failure Mode and Effects Analysis	A hazard identification technique in which all known failure modes of components or features of a system are considered in turn, and undesired outcomes are noted.
Fault Tree Analysis	A method used to analyze graphically the failure logic of a given event, to identify various failure scenarios (called cut-sets), and to support the probabilistic estimation of the frequency of the event.
F-N Curve	A plot of cumulative frequency versus consequences (often expressed as number of fatalities).
Flammable Liquids	Any liquid that has a closed-cup flash point below 100 °F (37.8 °C), as determined by the test procedures described in NFPA 30 and a Reid vapor pressure not exceeding 40 psia (2068.6 mm Hg) at 100°F (37.8 °C), as determined by ASTM D 323, Standard Method of Test for Vapor Pressure of Petroleum Products (Reid Method). Class IA liquids shall include those liquids that have flash points below 73 °F (22.8 °C) and boiling points below 100 F (37.8 °C). Class IB liquids shall include those liquids that have flash points below 73°F (22.8 °C) and boiling points at or above 100 °F (37.8 °C). Class IC liquids shall include those liquids that have flash points at or above 73 °F (22.8 °C), but below 100 °F (37.8 °C). (NFPA 30).
Frequency	Number of occurrences of an event per unit time (e.g., 1 event in 1000 yr. = 1×10^{-3} events/yr.).
Frequency Modeling	Development of numerical estimates of the likelihood of an event occurring.
Hazard	An inherent chemical or physical characteristic that has the potential for causing damage to people, property, or the environment.
Hazard Analysis	The identification of undesired events that lead to the materialization of a hazard, the analysis of the mechanisms by which these undesired events could occur and usually the estimation of the consequences.

Hazard and Operability Study (HAZOP)	A systematic qualitative technique to identify process hazards and potential operating problems using a series of guide words to study process deviations. A HAZOP is used to question every part of a process to discover what deviations from the intention of the design can occur and what their causes and consequences may be. This is done systematically by applying suitable guide words. This is a systematic detailed review technique, for both batch and continuous plants, which can be applied to new or existing processes to identify hazards
Hazard Identification	The inventorying of material, system, process and plant characteristics that can produce undesirable consequences through the occurrence of an incident.
Hazard Identification and Risk Analysis (HIRA)	A collective term that encompasses all activities involved in identifying hazards and evaluating risk at facilities, throughout their life cycle, to make certain that risks to employees, the public, or the environment are consistently controlled within the organization's risk tolerance.
Hot Work	Any operation that uses flames or can produce sparks (e.g., welding).
Impact	A measure of the ultimate loss and harm of a loss event. Impact may be expressed in terms of numbers of injuries and/or fatalities, extent of environmental damage and/or magnitude of losses such as property damage, material loss, lost production, market share loss, and recovery costs.
Inertion	A technique by which a combustible mixture is rendered non-ignitable by addition of an inert gas or a noncombustible dust.
Incident	An event, or series of events, resulting in one or more undesirable consequences, such as harm to people, damage to the environment, or asset/business losses. Such events include fires, explosions, releases of toxic or otherwise harmful substances, and so forth.

Independent Protection Layer (IPL)	A device, system, or action that is capable of preventing a scenario from proceeding to the undesired consequence without being adversely affected by the initiating event or the action of any other protection layer associated with the scenario. A protection layer meets the requirements of being an IPL when it is designed and managed to achieve the following seven core attributes: Independent; Functional; Integrity; Reliable; Validated, Maintained and Audited; Access Security; and Management of Change
Individual Risk	The risk to a person in the vicinity of a hazard. This includes the nature of the injury to the individual, the likelihood of the injury occurring, and the time period over which the injury might occur.
Inherent Safety	A condition in which the hazards associated with the materials and operations used in the process have been reduced or eliminated, and this reduction or elimination is permanent and inseparable from the process. Inherently safer technology (IST) is also used interchangeably with inherent safety in the book.
Inherently Safer Design	A way of thinking about the design of chemical processes and plants that focuses on the elimination or reduction of hazards, rather than on their management and control.
Interlock	A protective response which is initiated by an out-of-limit process condition. Instrument which will not allow one part of a process to function unless another part is functioning. A device such as a switch that prevents a piece of equipment from operating when a hazard exists. To join two parts together in such a way that they remain rigidly attached to each other solely by physical interference. A device to prove the physical state of a required condition and to furnish that proof to the primary safety control circuit.
Layer of Protection Analysis (LOPA)	An approach that analyzes one incident scenario (cause-consequence pair) at a time, using predefined values for the initiating event frequency, independent protection layer failure probabilities, and consequence severity, in order to compare a scenario risk estimate to risk criteria for determining where additional risk reduction or more detailed analysis is needed. Scenarios are identified elsewhere, typically using a scenario-based hazard evaluation procedure such as a HAZOP Study.

Likelihood	A measure of the expected probability or frequency of occurrence of an event. This may be expressed as an event frequency (e.g., events per year), a probability of occurrence during a time interval (e.g., annual probability) or a conditional probability (e.g., probability of occurrence, given that a precursor event has occurred).
Management of Change (MOC)	A system to identify, review and approve all modifications to equipment, procedures, raw materials and processing conditions, other than "replacement in kind," prior to implementation.
Management System	A formally established set of activities designed to produce specific results in a consistent manner on a sustainable basis.
Mechanical Integrity	A management system focused on ensuring that equipment is designed, installed, and maintained to perform the desired function.
Near-Miss	An unplanned sequence of events that could have caused harm or loss if conditions were different or were allowed to progress, but actually did not.
Off-Site Population	People, property, or the environment located outside of the site property line that may be impacted by an on-site incident.
Operating Procedures	Written, step-by-step instructions and information necessary to operate equipment, compiled in one document including operating instructions, process descriptions, operating limits, chemical hazards, and safety equipment requirements.
Organizational Change	Any change in position or responsibility within an organization or any change to an organizational policy or procedure that affects process safety.
Organizational Change Management (OCM)	A method of examining proposed changes in the structure or organization of a company (or unit thereof) to determine whether they may pose a threat to employee or contractor health and safety, the environment, or the surrounding populace.
OSHA Process Safety Management (OSHA PSM)	A U.S. regulatory standard that requires use of a 14-element management system to help prevent or mitigate the effects of catastrophic releases of chemicals or energy from processes covered by the regulations 49 CFR 1910.119.

Pre-Startup Safety Review (PSSR)

A systematic and thorough check of a process prior to the introduction of a highly hazardous chemical to a process. The PSSR must confirm the following: Construction and equipment are in accordance with design specifications; Safety, operating, maintenance, and emergency procedures are in place and are adequate; A process hazard analysis has been performed for new facilities and recommendations and have been resolved or implemented before startup, and modified facilities meet the management of change requirements; and training of each employee involved in operating a process has been completed.

Preventive Maintenance

Maintenance that seeks to reduce the frequency and severity of unplanned shutdowns by establishing a fixed schedule of routine inspection and repairs.

Probit

A random variable with a mean of 5 and a variance of 1, which is used in various effect models. Probit-based models derived from experimental dose-response data, are often used to estimate the health effect that might result based upon the intensity and duration of an exposure to a harmful substance or condition (e.g., exposure to a toxic atmosphere, or a thermal radiation exposure).

Process Hazard Analysis

An organized effort to identify and evaluate hazards associated with processes and operations to enable their control. This review normally involves the use of qualitative techniques to identify and assess the significance of hazards. Conclusions and appropriate recommendations are developed. Occasionally, quantitative methods are used to help prioritize risk reduction.

Process Knowledge Management

A Process Safety Management (PSM) program element that includes work activities to gather, organize, maintain, and provide information to other PSM program elements. Process safety knowledge primarily consists of written documents such as hazard information, process technology information, and equipment-specific information. Process safety knowledge is the product of this PSM element.

Process Safety Culture

The common set of values, behaviors, and norms at all levels in a facility or in the wider organization that affect process safety.

Process Safety Incident/Event	An event that is potentially catastrophic, i.e., an event involving the release/loss of containment of hazardous materials that can result in large-scale health and environmental consequences.
Process Safety Information (PSI)	Physical, chemical, and toxicological information related to the chemicals, process, and equipment. It is used to document the configuration of a process, its characteristics, its limitations, and as data for process hazard analyses.
Process Safety Management (PSM)	A management system that is focused on prevention of, preparedness for, mitigation of, response to, and restoration from catastrophic releases of chemicals or energy from a process associated with a facility.
Process Safety Management Systems	Comprehensive sets of policies, procedures, and practices designed to ensure that barriers to episodic incidents are in place, in use, and effective.
Qualitative Risk Analysis	Based primarily on description and comparison using historical experience and engineering judgment, with little quantification of the hazards, consequences, likelihood, or level of risk.
Quantitative Risk Analysis (QRA)	The systematic development of numerical estimates of the expected frequency and severity of potential incidents associated with a facility or operation based on engineering evaluation and mathematical techniques.
Reactive Chemical	A substance that can pose a chemical reactivity hazard by readily oxidizing in air without an ignition source (spontaneously combustible or peroxide forming), initiating or promoting combustion in other materials (oxidizer), reacting with water, or self-reacting (polymerizing, decomposing or rearranging). Initiation of the reaction can be spontaneous, by energy input such as thermal or mechanical energy, or by catalytic action increasing the reaction rate.

Recognized and Generally Accepted Good Engineering Practice (RAGAGEP)	A term originally used by OSHA, stems from the selection and application of appropriate engineering, operating, and maintenance knowledge when designing, operating and maintaining chemical facilities with the purpose of ensuring safety and preventing process safety incidents.

It involves the application of engineering, operating or maintenance activities derived from engineering knowledge and industry experience based upon the evaluation and analyses of appropriate internal and external standards, applicable codes, technical reports, guidance, or recommended practices or documents of a similar nature. RAGAGEP can be derived from singular or multiple sources and will vary based upon individual facility processes, materials, service, and other engineering considerations. |
Responsible Care©	An initiative implemented by the Chemical Manufacturers Association (CMA) in 1988 to assist in leading chemical processing industry companies in ethical ways that increasingly benefit society, the economy and the environment while adhering to ten key principles.
Risk	A measure of human injury, environmental damage, or economic loss in terms of both the incident likelihood and the magnitude of the loss or injury. A simplified version of this relationship expresses risk as the product of the likelihood and the consequences (i.e., Risk = Consequence x Likelihood) of an incident.
Risk Contour	Lines that connect points of equal risk around the facility ("isorisk" lines).
Risk Management Program (RMP) Rule	EPA's accidental release prevention Rule, which requires covered facilities to prepare, submit, and implement a risk management plan.
Risk Matrix	A tabular approach for presenting risk tolerance criteria, typically involving graduated scales of incident likelihood on the Y-axis and incident consequences on the X-Axis. Each cell in the table (at intersecting values of incident likelihood and incident consequences) represents a particular level of risk.

Risk-Based Process Safety (RBPS)	The Center for Chemical Process Safety's (CCPS) PSM system approach that uses risk-based strategies and implementation tactics that are commensurate with the risk-based need for process safety activities, availability of resources, and existing process safety culture to design, correct, and improve process safety management activities.
Risk Tolerance	The maximum level of risk of a particular technical process or activity that an individual or organization accepts to acquire the benefits of the process or activity.
Risk Tolerance Criteria	A predetermined measure of risk used to aid decisions about whether further efforts to reduce the risk are warranted.
Runaway Reactions	A thermally unstable reaction system which exhibits an uncontrolled accelerating rate of reaction leading to rapid increases in temperature and pressure.
Safety Instrumented Functions (SIF)	A system composed of sensors, logic servers, and final control elements for the purpose of taking the process to a safe state when predetermined conditions are violated.
Safety Instrumented System (SIS)	A separate and independent combination of sensors, logic solvers, final elements, and support systems that are designed and managed to achieve a specified safety integrity level. A SIS may implement one or more Safety Instrumented Functions (SIFs).
Safeguards or Protective Features	Any device, system, or action that either interrupts the chain of events following an initiating event or that mitigates the consequences. A safeguard can be an engineered system or an administrative control. Not all safeguards meet the requirements of an IPL.
Scenario	A detailed description of an unplanned event or incident sequence that results in a loss event and its associated impacts, including the success or failure of safeguards involved in the incident sequence.
Semi-Quantitative Risk Analysis	Risk analysis methodology that includes some degree of quantification of consequence, likelihood, and/or risk level.
Societal Risk	A measure of risk to a group of people. It is most often expressed in terms of the frequency distribution of multiple casualty events.

Standards

The PSM program element, Compliance with Standards, that helps identify, develop, acquire, evaluate, disseminate, and provide access to applicable standards, codes, regulations, and laws that affect a facility and/or the process safety requirements applicable to a facility. More generally, standards also refer to requirements promulgated by regulators, professional or industry-sponsored organizations, companies, or other groups that apply to the design and implementation of management systems, design and operation of process equipment, or similar activities.

Threshold Limit Value (TLV)

The maximum exposure concentration recommended by the American Conference of Government Industrial Hygienists (ACGIH) for long term exposures.

Threshold Limit Value-Time-Weighted Average (TLV-TWA)

The time-weighted average concentration limit for a normal 8-hour workday and a 40-hour workweek to which nearly all workers may be repeatedly exposed, day after day, without adverse effect. Developed by the ACGIH.

Toxicity

The quality, state, or degree to which a substance is poisonous and/or may chemically produce an injurious or deadly effect upon introduction into a living organism.

ACKNOWLEDGEMENTS

The Chemical Center for Process Safety (CCPS) thanks all of the members of the Guide for Making Acute Risk Decisions Subcommittee for providing technical guidance in the preparation of this book. CCPS also expresses its appreciation to the members of the Technical Steering Committee for their advice and support.

The co-chairs of the Subcommittee were Fred Henselwood of Nova Chemicals and Jeff Stawicki of Lubrizol. The CCPS staff consultant was David Belonger.

The Subcommittee had the following key contributing members:

Christopher Buehler	Exponent
Sorin Dan	Nova Chemicals
Elizabeth Lutostansky	Air Products
Robin Pitblado	DNV GL, (retired)
Martin Timm	Praxair
Florine Vincik	BASF

The following members also supported this project:

Seshu Dharavaram (Corteva); Derek Miller (Air Products); John Traynor (Evonik); Eric Peterson (MMI Engineering);

The collective industrial experience and know-how of the subcommittee members plus these individuals makes this book especially valuable to engineers who develop and manage process safety programs and management systems, including the identification of the competencies needed to create and maintain these systems.

The book committee wishes to express their appreciation to Albert Ness of CCPS his contributions in writing this book for publication.

Before publication, all CCPS books are subjected to a thorough peer review process. CCPS gratefully acknowledges the thoughtful comments and suggestions of the peer reviewers. Their work enhanced the accuracy and clarity of these guidelines.

Peer Reviewers:

Anne Bartelsman	Marathon Petroleum
Denise Chastain-Knight	Exida
Palaniappan Chidambaram	DuPont
Christopher F. Conlan	National Grid
Georges Melham	ioMosaic

Although the peer reviewers have provided many constructive comments and suggestions, they were not asked to endorse this book and were not shown the final manuscript before its release.

PREFACE

The Center for Chemical Process Safety (CCPS) was created by the AIChE in 1985 after the chemical disasters in Mexico City, Mexico, and Bhopal, India. The CCPS is chartered to develop and disseminate technical information for use in the prevention of major chemical accidents. The Center is supported by more than 200 chemical process industries (CPI) sponsors who provide the necessary funding and professional guidance to its technical committees. The major product of CCPS activities has been a series of guidelines to assist those implementing various elements of a process safety and risk management system. This book is part of that series.

The AIChE has been closely involved with process safety and loss control issues in the chemical and allied industries for more than five decades. Through its strong ties with process designers, constructors, operators, safety professionals, and members of academia, AIChE has enhanced communications and fostered continuous improvement of the industry's high safety standards. AIChE publications and symposia have become information resources for those devoted to process safety and environmental protection.

The integration of process safety into the engineering curricula is an ongoing goal of the CCPS. To this end, CCPS created the Safety and Chemical Engineering Education (SACHE) committee which develops training modules for process safety. One textbook covering the technical aspects of process safety for students already exists; however, there is no textbook covering the concepts of process safety management and the need for process safety for students. The CCPS Technical Steering Committee initiated the creation of this book to assist colleges and universities in meeting this challenge and to aid Chemical Engineering programs in meeting recent accreditation requirements for including process safety into the chemical engineering curricula.

1

INTRODUCTION

1.1 HISTORY OF APPROACHES TO PROCESS SAFETY MANAGEMENT

By the 1980's, incidents such as the Flixborough (UK) Nypro Plant explosion in 1974, the Seveso (Italy) dioxin release in 1976, the Piper Alpha oil platform explosion in 1976, the Mexico City (Mexico) LPG BLEVE explosion event, and the Bhopal (India) methylisocyanate release (both in in 1984) led to the development and promulgation of regulations concerning process safety throughout the world. A few examples are the:

- European Union Seveso Directives, (Seveso I, 1976, Seveso II, 1997 and Seveso III, 2012)
- U.S. OSHA Process Safety Management (PSM) rule, 1992
- U.K. Offshore Installations (Safety Case) Regulations 1992
- U.S. EPA Risk Management Plan (RMP) rule, 1997
- Mexican Integral Security and Environmental Management System (SISPA), 1998
- U.K Control of Major Accident Hazards (COMAH) regulation, 1999 (COMAH is the UK implementation of the Seveso directives)
- Canadian Environmental Protection Act, 1999
- Decree 591 (China): Regulations on Safe Management of Hazardous Chemicals, 2011

One problem with some of the regulations is that they are frequently limited in scope. For example, OSHA's PSM and the COMAH regulation have lists of covered chemicals with threshold quantities. Companies can choose to not apply any of the regulatory management system elements to non-covered chemical processes. The OSHA PSM rule exempts atmospheric storage tanks, yet such tanks have been involved in very serious incidents. Examples include the explosions in Buncefield, U.K. in 2005, and CAPECO (Caribbean Petroleum Company) in Puerto Rico in 2009, both due to overflow of atmospheric storage tanks.

In response to these regulations and public pressure, third party organizations also developed process safety guidance. In the U.S., American Petroleum Institute (API) published *Management of Process Hazards* in 1990 (API 750). The Center for Chemical Process Safety (CCPS) was established in 1985 in response to the Bhopal incident. The CCPS published *Guidelines for Implementing Process Safety Management Systems* (CCPS 1994) which had 15 process safety management system elements, and Guidelines for Risk Based Process Safety (CCPS 2007) which expanded this to twenty process safety management system elements. In Canada, the Canadian Chemical Producers Association (now the Chemical Industry Association of Canada, CIAC) developed Responsible Care® codes with its own management system elements (Topalovic and Krantzberg, 2014). The CIAC updated these codes in 2010.

A by no means inclusive list of other third-party organizations that have developed codes or support for process safety include:

- The International Association of Oil and Gas Producers (https://www.iogp.org/oil-and-gas-safety/process-safety/)
- The American Petroleum Institute (www.api.org) (https://www.api.org/news-policy-and-issues/safety-and-system-integrity)
- The Energy Institute (https://publishing.energyinst.org/topics/process-safety)
- Society of Organic Chemical Manufacturers Association http://www.socma.com/ChemStewards

CCPS now has over 200 member companies in countries on every continent (excluding Antarctica). The Responsible Care® codes were adopted not only by the U.S. Chemical Manufacturers Association (now the American Chemical Council (ACC)), but also in over 50 other countries. The various elements in these codes were based on industry experience and represented what process safety experts thought were the key elements of process safety management (PSM) at the time. With the worldwide reach of CCPS and Responsible Care®, these are truly internationally recognized.

1.2   THE   PARADIGM   OF   RISK-BASED   PROCESS   SAFETY MANAGEMENT

1.2.1 Risk Based Process Safety (RBPS) Management

Although the rules and regulations mentioned above are credited with improving process safety, many companies were "challenged with inadequate management system performance, resource pressures, and stagnant process safety results." (CCPS 2005, p ii).

Companies tended to adopt either standards based, regulations based, continuous improvement based, or some combination of these programs as their management system model. The shortcomings with these approaches are:

- Standards and regulations do not cover every situation
- Standards and regulations may represent the minimum that needs to be done to control risk
- Continuous improvement needs. If a company only looks at "lagging indicators", i.e., actual incidents in which people have

been hurt or killed, this could be insufficient, as they are (hopefully) rare.

> Adopting an approach that looks at "leading indicators" can enable continuous improvement. The reader should see the CCPS document, *Process Safety Leading and Lagging Metrics, You Don't Improve What You Don't Measure*, 2nd Ed., AIChE Industry Technology Alliance, Jan 2011, for guidance.
>
> (https://www.aiche.org/ccps/resources/tools/process-safety-metrics/references/Guidance-Documents)

In the early 2000's CCPS developed and published *Guidelines for Risk Based Process Safety* (RBPS) (CCPS 2007) to move to the next level of process safety management. Until then, the codes and regulations were hazards based, i.e., defined covered processes (usually by threshold levels of covered chemicals) and then had performance-based requirements for controlling the hazards. A risk-based approach to PSM recognizes that not all hazards are equal and emphasizes that the resources devoted to PSM should be appropriate to the hazards *and* risk of a given operation. RBPS also added several elements to process safety management. Table 1 compares the elements of RBPS, the original CCPS PSM elements, and the Responsible Care® Code practices. Space considerations prevent a listing of elements from the many other international regulations and third-party codes; however, they have elements that are similar to these.

Adopting a risk-based approach in addition to regulatory compliance, adherence to code and standards and good engineering practices means that decisions about process safety (e.g., how much risk reduction is needed, or which process option, safety control, shutdown system should be installed) are now risk-based decisions. They explicitly consider consequences and likelihood, and they are not based solely on hazards (with the provision that applicable regulations are also followed) whether an organization realizes it or not. The definitions of hazard, consequence, likelihood and risk are:

- **Hazard**. An inherent chemical or physical characteristic that has the potential for causing damage to people, property, or the environment.
- **Consequence**. The undesirable result of a loss event, usually measured in health and safety effects, environmental impacts, loss of property, and business interruption costs.
- **Likelihood**. A measure of the expected probability or frequency of occurrence of an event. This may be expressed as an event frequency (e.g., events per year), a probability of occurrence during a time interval (e.g., annual probability) or a conditional probability (e.g., probability of occurrence, given that a precursor event has occurred).

Table 1. Comparison of RBPS, Original CCPS PSM Elements, and Responsible Care® Code Practices

RBPS 2005	CCPS 1994	Responsible Care® 2010
Process Safety Culture	Accountability: Objectives and Goals	Leadership and Culture
Compliance with Standards	Standards, Codes, and Laws	Accountability
Process Safety Competency	Enhancement of Process Safety Knowledge	Knowledge, Expertise and Training
Workforce Involvement		Information Sharing
Stakeholder Outreach	Process Knowledge and Documentation	Understanding and Prioritization of Process Safety Risks
Process Knowledge Management	Capital Project Review and Design Procedures	Comprehensive Process Safety Management Systems
Hazard Identification and Risk Assessment	Process Risk Management	Monitoring and Improving Performance
Operating Procedures	Training and Performance	
Safe Work Practices	Human Factors	
Asset Integrity and Reliability	Process and Equipment Integrity	
Contractor Management	Management of Change	
Training and Performance Assurance	Incident Investigation	
Management of Change	Audits and Corrective Actions	
Operational Readiness		
Conduct of Operations		
Emergency Management		
Incident Investigation		
Measurement and Metrics		
Auditing		
Management Review and Continuous Improvement		

- **Risk**. A measure of human injury, environmental damage, or economic loss in terms of both the incident likelihood and the magnitude of the loss or injury. A simplified version of this relationship expresses risk as the product of the likelihood and the consequences (i.e., Risk = Consequence x Likelihood) of an incident.

CCPS published *Tools for Making Acute Risk Decision with Chemical Process Safety Applications* (CCPS 1995), which focused on decision aids such as, voting, weighted scoring, and other, more complicated, methods, such as cost-benefit analysis. The audience for that book was mainly decision makers and risk analysts. Some of the tools described involved sophisticated mathematics.

The decision tools referred to in the 1995 *Tools for Acute Risk Decision* book will be called decision aids, and the term decision tools will refer to risk analysis tools used to evaluate decision alternatives. This book is titled Practical "Guide to Making Acute Risk Decisions" and is therefore not a second edition about making risk decisions. Instead, it is a companion book, hence the change in the title. This book is meant to be a guide to the decision process in general and to the use of common and practical risk evaluation and risk analysis tools to analyze decision alternatives. The intended audience is the decision teams who analyze the options and the people who make process safety risk decisions. That would include engineers who design and run plants, chemists and engineers who do conduct process research, Health, Safety, and Environmental (HSE) professionals, plant and engineering managers, and business managers.

1.2.2 Risk Decisions Characteristics

Proportionality. First and foremost, risk-based process safety and decision making recognizes that all risks associated with hazards are not equal. Although no risk should go uncontrolled, an organization always needs to allocate resources efficiently and effectively. This requires identifying the hazards for all processes, understanding the potential impacts, and assigning appropriate resources to control them. For example, a well understood hazard and risk may be adequately controlled by applying the recommendations of a consensus standard or developing an internal design and operating standard. A complex, less well understood hazard may need multidisciplinary team to analyze the risk and define safeguards. Organizations trying to implement RBPS management may turn to the use of risk criteria to aid in determining the resources and level of safeguards needed.

Competing objectives. Risk decisions are not made in isolation; there are always tradeoffs. Process safety risk decisions are made in conjunction with consideration of factors such as; capital and operating costs, resource availability and allocation, additional complexity of new safety systems (which increases maintenance costs and training costs), time frames, the regulatory environment, relationships with the local community, and business objectives and imperatives, to name some. Therefore, risk decisions frequently take many or all of these other factors into consideration. Obviously, a balance is needed. As a decision problem is defined, these other objectives need to be considered. Use of the appropriate decision aid, as described in Chapter 3, can help in this process.

Multiple stakeholder involvement. Stakeholder Outreach is one of the RBPS process safety elements. It is critical to engage relevant stakeholders who will be affected by the risks directly or indirectly, including those who are likely to influence decisions that will have an impact on effectiveness of risk management. Often, stakeholders considered are limited to operations and maintenance function in the organization while functions like procurement, supply chain, etc., are not considered as relevant. Unless a decision is a simple one, there can be more than one stakeholder. In addition to the stakeholders inside the organization, a decision can affect the local community. Sharing information with stakeholders can lead to a better final decision and builds trust with them

Uncertainty. The recommended approach to risk decisions throughout this book is to use the simplest analysis tools and decision aids possible and advance to more complex tools only if needed. Risk decisions require defining risk scenarios, data about frequencies and likelihoods of initiating events, consequences of material and/or energy releases, and failure probabilities of layers of protection. All of these items have margins of error that can impact the decision. Hence decision makers need to be aware of the uncertainty in the information to decide if more complex tools, and possibly more accurate information, are needed.

1.3 A RISK DECISION MAKING METHOD

Many examples of risk decision making processes can be found in literature. This book describes one approach to the decision-making process, based on a blend of the decision process in the 2005 CCPS decision tools book cited above, and a process described in *Smart Choices* (Hammond, Kenney, and Raiffa, 1999). The key elements are essentially the same in all the processes, although the number and names of steps may be different.

The resulting risk decision process for this book is:

1. Define the problem
2. Evaluate the baseline risk
3. Identify the alternatives
4. Evaluate the alternatives
5. Make the decision

Taking time to define the problem will ensure the right problem is being studied and that the level of complexity is understood. Defining the problem includes defining the objectives and constraints. The objectives may be considered first to aid in problem definition. Thinking about objectives of the decision and desired outcomes will provide direction to your efforts.

Without alternatives there is no decision, however, all too often an organization may only consider alternatives that it is familiar with. More and better options can be defined if a team of people takes the time to think about them. Thinking about the objectives of the decision can help an organization identify more creative options. An unidentified option can't be evaluated.

When evaluating options, one should start with the simplest tools available. If an organization properly defines the problem and objectives, identify good options, and use the right tools to evaluate them, the chances of making a better decision increase.

The decision aid is tailored to the complexity of the decision. A simple decision can be made by weighing pros and cons. A complicated decision involving many trade-offs will need a more sophisticated decision aid, such as a weighted scoring method.

1.4 ROAD MAP AND RELATIONSHIP OF THIS BOOK WITH OTHER MATERIAL

Figure 1.1 provides a road map through this book. The reader can use this to focus in on whatever topic they feel is important to them.

Chapters 2 and 3 are a brief review of risk analysis and process hazards. They are intended for those who are not process safety subject matter experts and need a little education on these subjects. Process safety subject matter experts may find these chapters are good refreshers.

All audiences should read Chapters 4, 5, and 6:

- Chapter 4 outlines the decision process and aids
- Chapter 5 characterizes decision complexity and selection of risk decision tools
- Chapter 6 describes common decision traps.

Even if your organization has its own decision process and tools, Chapters 4 and 5 may help to improve, refine or just make the decision process more organized. Regardless of your knowledge and sophistication about risk decision making, Chapter 6, Decision Traps, should be required reading for anyone making any decisions. This chapter

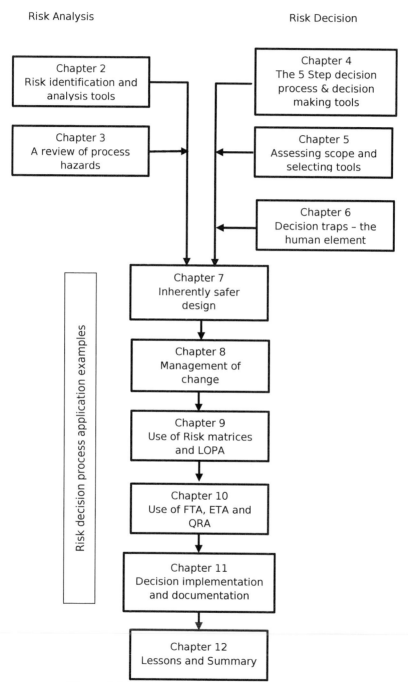

Figure 1.1. Risk Decision and Risk Management Road Map

outlines how our own very human tendencies and failings can lead to poor decisions.

Chapters 7 through 10 provide examples of how the decision process and tools can be applied and are an aid for people new to risk decision making. Process safety subject matter experts may also find them beneficial.

Chapter 11 describes documentation and implementation of a decision. Again, process safety experts may find this to be a familiar story, but all others should read this, as proper documentation and dissemination of process safety knowledge is a key element of RBPS.

Chapter 12 contains examples of how poor decision making has led to process safety incidents. These examples can help you convince people of the need for a systematic approach to making risk decisions and are useful for all audiences.

1.5 RISK DECISIONS DURING PROCESS LIFE CYCLE

In *Inherently Safer Chemical Process: A Life Cycle Approach* (CCPS 2009a) the stages of a process's life were described as:

- Research and Development
- Process Development
- Process Design
- Operations, Maintenance, Management of Change
- Decommissioning and Abandonment

Decisions about process safety early in the life cycle can be quick/simple decisions, that may have enormous consequences to subsequent life cycle stages. For example, as an individual, a research chemist might have to decide between two solvents to run a reaction in. The chemist can make that decision as an individual. Although it seems obvious to a process engineer or process safety expert that choosing solvents with higher flashpoints or lower toxicity (e.g., let's use water instead of toluene) is preferable, the chemist might consider only which solvent is the best for the reaction (the reaction is faster and more selective if run in toluene). Long term costs to the organization may be incurred to manage the risk from the decision of the chemist.

That same decision, during process development, can still be reversed at little cost. However, if postponed to process design, the decision could lead to delays if the organization wants to change the solvent. By the time the process goes into operation, it might be very difficult or costly to make a change. Such a decision at a later stage becomes a complex one, involving the business and operations as well as safety.

The key here is to train personnel to understand process safety issues and to recognize when a decision affects them. This is part of the intent of the Process Safety Competency element of RBPS.

1.6 PROS AND CONS

Pros. As mentioned above, a risk-based approach goes beyond a hazards/regulatory based approach and looks at probability and consequences as well. A risk-based approach enables more effective allocation of resources. Inclusion of a risk criteria provides a measure to show when enough risk reduction has been achieved as well as to ensure risk-based decisions are consistent.

Cons. A company can use risk analysis tools that are more complicated than needed for a decision, hence wasting resources. Worse yet, a company can fall victim to the "paralysis by analysis" syndrome, over analyzing a problem and thus postpone or not act. There may also be a problem of explaining risk to management and stakeholders, especially for High Consequence Low Probability (HCLP) events. It may seem easier to them to want to ignore that such low probability risks even exist. It is critical to ensure organization and especially leadership understands the fundamentals of risk and risk-based approach to lend its support in the process. All stakeholders, however, may not be familiar with the concept of risk and discussing risk, and may be troubled by the concept of a tolerable risk.

One way to address HCLP events, a company might adopt a consequence-based versus risk-based approach. Addressing HCLP events using a risk-based criterion will be discussed more in Section 10.3 – *High Consequence Low Probability Events.*

1.7 SUMMARY

The concept of risk-based process safety (RBPS) was developed by the CCPS to improve process safety performance.   RBPS provides a framework for process safety efforts to move beyond compliance with standards and regulations, and a limited hazards-based approach, by doing what is needed based on process risks.

Whereas the 1995 book on decision tools focused on the aids for making the final choice, this book looks at the overall decision process. Decision tools described in the 1995 book are referred to as decision aids in this book. The concept and process of risk management is explained, and a five step decision making process (developed based on a combination of the original process in the 1995 decision tools book and the one described in *Smart Choices* (Hammond, Kenney, and Raiffa, 1999) is outlined; define the problem, evaluate the baseline risk, identify alternatives, evaluate the alternatives, and make the decision.

This decision process is then illustrated with examples of real problems that have been described in open literature. The application of qualitative and quantitative risk analysis tools such as risk matrices, layer of protection analysis, fault tree analysis, event tree analysis and quantitative risk analysis is demonstrated in these examples. The concept of risk criteria is briefly explained to allow you to see how risk criteria fit into the risk decision process.

2

KEY CONCEPTS IN RISK MANAGEMENT

2.1 RISK MANAGEMENT PROCESS

This chapter presents an outline of the Risk Management process. The basis for this outline comes from ISO 31000, *Risk Management – Principles and Guidelines* (ISO 2009), which identifies four steps in the risk assessment process:

1. *Risk identification* (also referred to as scenario identification) in which the sources of risk and their impacts are identified (ISO 2009, Section 5.3).
2. *Risk analysis*, in which the likelihood and magnitude of the risk is established.
3. *Risk evaluation*, in which the result of the risk analysis is compared to the organization's risk tolerance criteria.
4. *Risk treatment,* in which the options for risk reduction are chosen.

These steps were illustrated in Figure 2.1. Steps 1 through 3 are briefly described in this chapter. Steps 2 and 3 are described more fully in Chapters 9 and 10. Step 4, choosing the risk reduction options, is the subject of Chapter 4 and is also illustrated in Chapters 7 through 10. Chapters 2 and 7 through 10 together make up risk-based process safety element Hazard Identification and Risk Analysis (HIRA).

Prior to any step is setting the scope of the risk study, which is not covered in this book. After a risk management study is completed and the findings addressed, management then needs to ensure that the risk level is maintained as the facility is operated. Although this is not within in the scope of this book, Chapter 11 – *Good Decision Making and Implementation* does address good documentation and revalidation, things needed to enable good risk assurance.

2.2 RISK IDENTIFICATION – RISK SCENARIO

A Process Safety risk scenario is a detailed description of an unplanned event or incident sequence that results in a loss event and its associated impacts, including the success or failure of safeguards involved in the incident sequence. In a perfect world, all risk scenarios would be

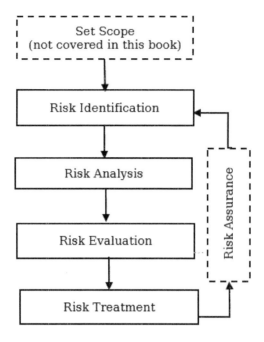

Figure 2.1. Risk Assessment Process (Adapted from ISO 3100)

identified in a Process Hazard Analysis (PHA) (e.g. checklist, what-if, HAZOP study). A PHA should address the following questions:

1. What are the hazards/what can go wrong?
2. How severe could it be (consequence)?
3. How likely is it to happen (frequency or likelihood)?
4. How do consequence and frequency combine? What is the risk associated with the hazardous event (risk)?
5. Is the current level of risk tolerable, considering existing layers of protection or risk controls?
6. If not, what needs to be done to reduce and manage the risk to an acceptable level?

In some US companies, PHA has come to specifically mean the study done to comply with the OSHA PSM standard. In this book PHA is meant to apply to any process hazard analysis, whether done for regulatory compliance or not.

Sadly, this is not a perfect world, PHAs can overlook scenarios (see Revalidating Process Hazard Analysis (CCPS 2001, p75-84) for checklists to evaluate PHAs) and many risk scenarios are identified after a near miss or incident has occurred. When this happens, a forward-looking company will not only look for ways to reduce the risk from the sequence of events that just occurred, but also from other events that could have caused a similar loss event. Also, the forward-looking company will study other loss events that could have occurred if safeguards that functioned properly had failed.

2.2.1 Risk Identification

There are several PHA methodologies that are applicable for identifying hazard and risk scenarios. Those described in *Guidelines for Hazard Evaluation Procedures, 3rd Ed.* (CCPS 2008) are:

What-If Analysis. The What-If analysis (and the What-If/Checklist) is a brainstorming technique. It is more unstructured than the other techniques listed here; however, by brainstorming a list of "What-If" questions (e.g. "what if the cooling system fails?" or "what-if too much chemical A is added?") a table of hazard scenarios can be constructed. Supplementing a What-If analysis with a checklist based on experience with the technology or process can improve the effectiveness of a What-If review.

Hazard and Operability (HAZOP) Study. HAZOP is sometimes called a structured brainstorming technique. In a HAZOP a process is systematically reviewed by applying guidewords to the process parameters to create process deviations (e.g. less cooling, higher pressure, or more chemical A). The review team then lists causes, consequences, safeguards and, if necessary, risk reduction recommendations. Each cause-consequence pair in the HAZOP is a hazard scenario. HAZOPs are frequently combined with risk matrices as described in Section 2.3.3 to do a qualitative estimate of risk.

Failure Modes and Effects Analysis (FMEA). For an FMEA a list of equipment items is developed. An equipment description, i.e. type, operating configuration, service characteristics is created for each item (e.g., "motor operated valve, normally open, in a three-inch sulfuric acid line). The team then lists all the failure modes, their effects, safeguards and, if necessary, risk reduction recommendations. Each failure mode-effect pair is a cause-consequence pair. FMEAs are also frequently combined with risk matrices to do a qualitative estimate of risk.

2.3 RISK ANALYSIS - CONSEQUENCES AND FREQUENCY

For process safety, risk is defined as a measure of human injury, environmental damage, or economic loss in terms of both the incident likelihood and the magnitude of the loss or injury. A simplified version of this relationship expresses risk as the product of the likelihood and the consequences (i.e., Risk = Consequence x Likelihood) for a risk scenario. Risk analysis involves determining the consequences of concern and estimating or calculating the frequency or likelihood of that consequence. This section describes these steps.

2.3.1 Consequences and Impacts

The loss event described in the scenario definition is the consequence or impact of the event. Defining the loss event and its consequences is step 2 of the PHA process. A loss event can be defined in several ways:

- The loss of containment of a hazardous material or energy, for example the loss of a specified amount of a toxic material or rupture of a vessel under pressure.
- The possible result of that release, for example, a fire, explosion overpressure, or toxic exposure.
- The ultimate impact of the loss event, such as an injury or fatality.

A *consequence-based criterion* may be established for any of these outcomes. These undesirable consequences or impacts can be defined to different levels of detail ranging from experiences from past events to using computational methods.

In the PHA, the consequences are usually assessed using experience and engineering judgement. Simple models are sometimes used to refine these estimates.

2.3.2 Frequency

Frequency is defined as the number of occurrences of an event per unit time (e.g., 1 event in 1,000 yr. = 1×10^{-3} events/yr.). Estimating the frequency of a scenario is step 3 of a PHA. The initiating event frequency is usually determined from operating experience, engineering judgment, or generic failure rate information with existing safeguards in place. Teams tend to be poor at qualitatively estimating the frequency of low probability events. Therefore, it is sometimes necessary to proceed to quantitative calculations to determine the frequency. In quantitative risk calculations, the scenario frequency is determined by multiplying the initiating event frequency by the probability of failure of the safeguards in place to prevent the consequence of concern. Tools for calculating frequency and probabilities are briefly described in Section 2.3.3.

2.3.3 Risk Estimation

Qualitative, Semi-Quantitative and Quantitative Risk. Risk estimates can be completely qualitative, i.e., based purely on the judgement of a hazard assessment team during or after an assessment. Risk estimates can be completely quantitative, as with a quantitative risk analysis (QRA). In a full QRA, risk is calculated using consequence models and frequency calculation tools such as a fault tree analysis (FTA). In between these two extremes there can be a mixture of judgement and calculations. For example, in a layer of protection analysis (LOPA), the consequence of concern may be determined by a team's judgment, and the frequency calculated. The intermediate methods are sometimes referred to as semi-quantitative. Some people do not use the term semi-quantitative, believing that any tool using numbers is quantitative, but the term semi-quantitative is widely used and understood by most practitioners in this field. Consequences and frequencies can be combined using the following risk tools, listed in order of increasing rigor of quantitative analysis:

- Engineering judgement
- Risk Matrices
- Risk Graphs
- Bow Tie Analysis
- Layer of Protection Analysis (LOPA)
- Fault Tree Analysis (FTA)
- Event Tree Analysis (ETA)
- Cause-Consequence Analysis
- Quantitative Risk Analysis (QRA).

As one progresses though these methods their precision increases, but this comes at the cost of increasing time and manpower. Conversely, as these methods become more complex, the reproducibility of the answers can decrease if an organization does not implement adequate procedures to guard against the problem of reproducibility.

Risk Matrices. CCPS defines a risk matrix as "A tabular approach for presenting risk tolerance criteria, typically involving graduated scales of incident likelihood on the Y-axis and incident consequences on the X-Axis. Each cell in the table (at intersecting values of incident likelihood and incident consequences) represents a particular level of risk". Risk matrices can be qualitative or semi-quantitative. Figure 2.2 and Tables 2.1, 2.2 and 2.3, from *Guidelines for Developing Quantitative Safety Risk Criteria* (CCPS 2009) shows a qualitative risk matrix. Tables 2.4 and 2.5 show a semi-quantitative risk matrix, also from CCPS (2009).

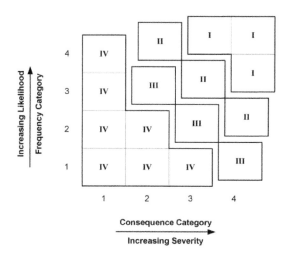

Figure 2.2. Qualitative Risk Matrix Example

Table 2.1. Example Consequence Categories – Qualitative Risk Matrix

Category	Description
1	No injury or health effects
2	Minor to moderate injury or health effects
3	Moderate to severe injury or health effects
4	Permanently disabling injury or fatality

Table 2.2. Example Frequency Categories – Qualitative Risk Matrix

Category	Description
1	Not expected to occur during life of process/facility
2	May occur once during life of process/facility
3	May occur several times during life of process/facility
4	Expected to occur more than once in a year

Table 2.3. Example Risk Ranking/Response Categories – Qualitative Risk Matrix

Risk Level	Description	Required Response
I	Unacceptable	Immediate mitigation or termination of activity
II	High	Mitigation within 6 months
III	Moderate	Mitigation within 12 months
IV	Acceptable As Is	No mitigation required

Table 2.4. Example Semi-Quantitative Risk Matrix [DOD 2000]

Severity Probability	Catastrophic	Critical	Marginal	Negligible
Frequent	1	3	7	13
Probable	2	5	9	16
Occasional	4	6	11	18
Remote	8	10	14	19
Improbable	12	15	17	20

TABLE 2.5. Suggested Mishap Severity Categories for Semi-Quantitative Risk Matrix [DOD 2000]

Description	Category	Environmental, Safety, and Health Results Criteria
Catastrophic	I	Could result in death, permanent total disability, loss exceeding $1M, or irreversible severe environmental damage that violates law or regulation.
Critical	II	Could result in permanent partial disability, injuries or occupational illness that may result in hospitalization of at least three personnel, loss exceeding $200K but less than $1M, or reversible environmental damage causing a violation of law or regulation.
Marginal	III	Could result in injury or occupational illness resulting in one or more lost work day(s), loss exceeding $10K but less than $200K, or mitigatible environmental damage without violation of law or regulation where restoration activities can be accomplished.
Negligible	IV	Could result in injury or illness not resulting in a lost work day, loss exceeding $2K but less than $10K, or minimal environmental damage not violating law or regulation.

CCPS (2009) provides more detail on qualitative and semi-quantitative risk matrices. During a PHA study, the PHA team can estimate the consequences and hence the consequence level of the scenario, usually based on judgement. Next, the team estimates the current frequency based on the likelihood of the consequence. Applying frequency and consequence ratings to causes (or failure modes) and consequences (or effects) during a PHA is commonly done by many chemical manufacturers.

Risk Graphs. Risk graphs are a semi quantitative tool, described in a German standard (DIN 1994), to estimate the Safety Integrity Level (SIL) of a Safety Instrumented Function (SIF) needed to obtain a needed level of risk reduction. Figure 2.3 shows a risk graph. The parameters used are:

- C = Consequence of the hazardous event
- F = Frequency of presence in the hazardous zone and the probability of exposure time, or occupancy
- P = Probability of avoiding the hazardous event
- W = Probability of the unwanted occurrence.

The risk graph parameters are further refined as:

- F1 = Rare to more frequent exposure in the hazardous zone
- F2 = Frequent to permanent exposure in the hazardous zone
- P1 = Possible under certain circumstances
- P2 = Almost impossible

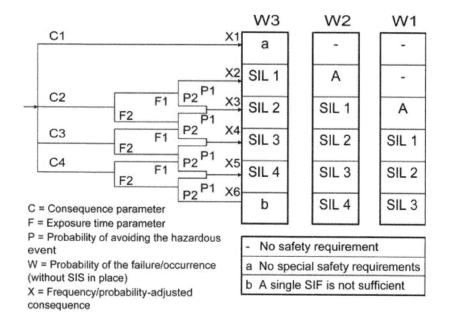

Figure 2.3. Example of risk graph modeled on IEC 615 11 (Baybutt, 2014)

Layer of Protection Analysis. Layer of Protection Analysis (LOPA) is a widely used form of semi-quantitative risk analysis. LOPA typically uses order of magnitude categories for the initiating event frequency, consequence, and the probability of the occurrence of enabling conditions, conditional modifiers, and failure of independent protection layers (IPLs), to analyze and assess the risk of one or more scenarios. The final scenario frequency is the product of the initiating event, the enabling conditions, the conditional modifiers, and the IPLs.

Some organizations use initiating event frequencies and IPL failure probabilities based on data and/or detailed quantitative analysis to refine the precision of a LOPA. For the remainder of this book, this will be called an enhanced LOPA.

The following CCPS books provide detailed information about performing a LOPA:

- LOPA Layer of Protection Analysis (CCPS 2001)
- Guidelines for Enabling Conditions and Conditional Modifiers in Layer of Protection Analysis (CCPS 2014)
- Guidelines for Initiating Events and Independent Protection Layers (CCPS 2015)
- Guidelines for Hazard Evaluation Procedures, 3rd Ed (CCPS, 2008)
- Guidelines for Chemical Process Quantitative Risk Analysis, 2nd Ed. (CCPS 1999).

Hybrid Methods. Baybutt (2007) proposed a modification to convert the risk graph into a method more aligned with Layer of Protection Analysis (LOPA). This will not be discussed here; the reader is referred to the reference for details.

Fault Tree Analysis (FTA). A fault tree is a graphical model that illustrates combinations of failures that will cause one specific consequence or failure of interest, called a Top event. FTA will be described in more detail in Chapter 11.

Event Tree Analysis (ETA). Event Tree Analysis is used to identify outcomes from an initiating event. ETA uses a graphical model that shows the possible outcomes following the success or failure of protective systems after an initiating event, such as loss of material or energy, has occurred.

Cause-Consequence Analysis (CCA). CCA combines the FTA and ETA into one diagram. CCA is useful for scenarios have relatively simple logic.

All these techniques are described in more detail in *Guidelines for Hazard Evaluation procedures, 3rd Edition* (CCPS 2008).

Quantitative Risk Analysis (QRA). Quantitative Risk Analysis develops numerical estimates of the expected frequency and severity of multiple possible scenarios associated with a facility or operation. The results are compared to the safety risk criteria.

This information is combined to calculate a numerical risk value. Typical risk measures are:

- *Risk Indices.* A risk index is a single number that show the magnitude of the risk. One example is the Fatal Accident Rate (FAR). This is the estimated number of fatalities per 10^8 exposure hours (roughly equal to the cumulative number of working hours over the lifetimes of 1,000 employees) (Lees, 2004).
- *Individual risk measures.* Individual risk is the risk of a single person exposed to the hazard.
- *Societal risk measures. Societal risk* expresses the risk to groups of people who might be affected by events that have the potential to affect large numbers of people. Societal risk is usually expressed as a risk of one or more fatalities per year.

QRA is described in more detail in Guidelines for Chemical Process Quantitative Risk Analysis, 2nd Edition (CCPS 1999). A more complete description of various risk measures, as well as a history of risk measures, is provided in the book *Guidelines for Developing Quantitative Risk Criteria* (CCPS, 2009).

2.4 RISK EVALUATION

2.4.1 Decision criteria

After the risk of a process or facility has been analyzed, the risk is evaluated by comparing it to criteria established by the organization. When setting risk tolerance criteria, some of the things an organization needs to consider include:

- The Company's risk appetite to accept risks
- Local and national regulations
- The magnitude of a consequence the organization is willing to accept and/or capable of managing
- Stakeholder expectations
- The organization's culture
- The company operations and technical capabilities
- How the risks will be analyzed
- How risk decisions will be made throughout the organization
- The training necessary for the workforce to understand the risk tolerance criteria

- How to implement risk decisions consistently throughout the organization
- Will the criteria be applied to individual risk scenarios, or to sum of all the scenarios identified in a unit or facility?
- Will the criteria be applied to new designs or to existing assets having the same risk?

Risk Tolerance. Risk tolerance describes how much risk an organization is willing to tolerate to achieve the desired benefits. Figure 2.4 shows a framework for discussing the tolerability of risk used by the Health and Safety Executive (HSE) in the UK (HSE, 2001). Figure 2.4 illustrates the concept that as risk increases, the tolerability of that risk decreases. There are three regions of risk shown in Figure 2.4:

- Broadly acceptable
- Tolerable
- Unacceptable

Each organization needs to establish these regions for itself. At the highest level in the organization lines need to be drawn between the Unacceptable and the Tolerable regions, and the Tolerable and Broadly Acceptable regions in Figure 2.4.

For example, in the risk matrix in Figure 2.2, Region IV would be equivalent to the broadly acceptable region, Region I would be equivalent to the unacceptable region, and Regions II and III are equivalent to increasing degrees of tolerability. In the definition of the categories in Table 2.3, Region II is tolerable only for a short time (six months), Region III for a longer time (one year).

The decision needs to be made at the highest levels in the organization because the line between the unacceptable and tolerable levels defines how much risk reduction a facility has to implement to prevent low probability - high consequence events that can have a major impact on the viability and reputation of the whole organization. Examples are events such as the Bhopal methylisocyanate release or BP Texas City refinery explosion. The line between the tolerable and broadly acceptable regions will tell a facility when no further risk reduction is needed.

Some companies may not have explicit risk tolerance criteria. Rather than defining an area on a risk matrix that is considered acceptable or unacceptable, they may use compliance with industry standards as acceptance criteria, or they may have approval levels (decision levels) defined based on risk or consequence. In either case, the decision to use such methods still needs to be made at the same levels as a decision to establish and use a risk tolerance criterion.

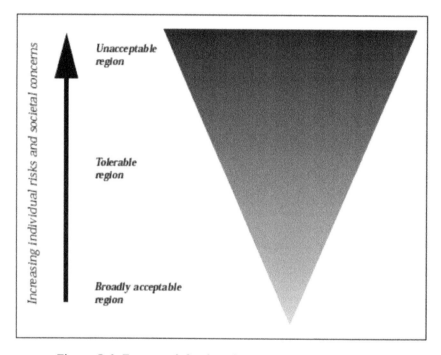

Figure 2.4. Framework for the tolerability of Risk (HSE, 2001).

As Low as Reasonably Practicable (ALARP). The concept of ALARP can be part of any risk tolerance criteria. With ALARP efforts to reduce risk should be continued until the incremental sacrifice (in terms of cost, time, effort, or other expenditure of resources) is grossly disproportionate to the incremental risk reduction achieved. The terms So Far As Is Reasonable Practicable (SFAIRP), and As Low as Reasonably Achievable (ALARA) are sometimes used in place of ALARP. In the UK, ALARP is used to describe the level to which risks in the workplace must be controlled.

An advantage of using ALARP as a risk decision criterion is simplicity, especially if combined with a qualitative risk criterion and a pre-defined risk reduction strategy. ALARP also provides a way to determine if an organization has done enough risk reduction when there is difficulty in meeting semi-quantitative or quantitative risk criteria.

The disadvantages of ALARP as a risk decision tool are:

- Expert judgment is needed to decide what is disproportionate to the incremental risk reduction when using ALARP as a risk decision tool.
- Formal techniques, such as cost-benefit analysis and the tools described below, may be needed for complex decisions or highly hazardous situations.

2.4.2 Qualitative, Semi-Quantitative and Quantitative Risk Criteria

When trying to determine tolerable risk criteria, an organization needs to choose whether to use qualitative, semi-quantitative, or quantitative risk criteria. Some countries have regulations establishing quantitative risk criteria. Again, the reader is referred to *Guidelines for Developing Quantitative Risk Criteria* (CCPS, 2009) for more details about determining risk criteria.

Qualitative Risk Criteria. If a company uses qualitative risk criteria, the risk assessment is most likely done using a qualitative risk matrix, such as the one as shown in Figure 2.2, during the hazard assessment. The team makes engineering judgments about how severe and how likely any given scenario is as the scenario is being discussed. Training about making these assessments of consequence and likelihood needs to be provided to the team before the PHA. Subject matter experts can help make these judgements.

Guidance needs to be provided to establish a risk reduction strategy with qualitative risk criteria. One example of risk reduction strategy that can be used with qualitative criteria is a guideline for the number of prevention and mitigation barriers for high, medium and low risk scenarios. With a combination of a qualitative risk matrix (see Section 2.3.3) and a hierarchy of risk reduction strategies (Figure 2.5, from Inherently Safer Chemical Process (CCPS 1996)) the person or team making the decision has guidance as to what options to select within the parameters of how much time and capital is available to implement the chosen option.

Even with guidance, judgments about consequences, frequency, and the tolerability of the resulting risk are still made on a subjective basis using the collective knowledge and experience of the PHA team members. The risk decision judgment may not be consistent from team to team across the plant or organization, or even within the same study conducted by the same team unless the company develops consistent training and competency for how to perform a risk assessment and uses some type of internal governance, assurance or quality control to review completed risk assessments.

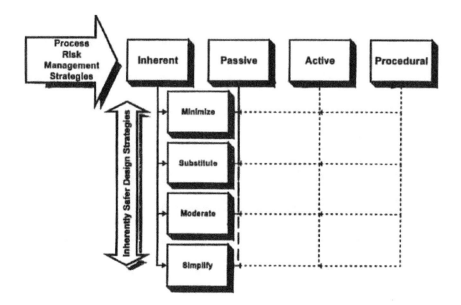

Figure 2.5. Hierarchy of risk reduction strategies (CCPS 1996)

Semi-Quantitative and Quantitative Risk Criteria. A semi-quantitative or quantitative risk tolerance criterion uses numbers for the frequency and/or the severity of the consequences. The criteria are typically expressed as a defined consequence per unit of time. Examples are:

- The likelihood of a release of 1,000 – 10,000 pounds of an extremely toxic material that can form a vapor at ambient conditions must be less than 1/1,000 per year.
- The likelihood of one or more fatalities on-site must be less than 1/10,000 per year.

The first example uses severity categories based on release sizes, whereas the second example uses the potential impact of the event, which could be a fire, explosion, or one or more fatalities. In a semi-quantitative risk analysis frequency and consequences are described in categories in which the frequencies and/or consequences have a numerical estimate associated with them. For example, a frequency category might be "occurring more than once during the life of the process or between 1/10 to 1/100 years", assuming a plant may run for 30 or 40 years.

If a company uses risk criteria, the risk assessment is likely to start with a semi-quantitative risk matrix, such as shown in Table 2.4, during the PHA.

In Section 1.3 – *A Risk Decision Making Method*, it was pointed out that an organization should start with the simplest tools first and move onto more complex ones as needed. As the risk analysis and evaluation proceeds, an organization can use the risk analysis tools described in Section 2.3 in order of increasing complexity and accuracy until the decision team believes it has enough information to make a good decision. For example, a study can begin by using a risk matrix in a PHA. Then a decision may be made to proceed to a LOPA. If the LOPA results do not provide the decision makers with enough information or detail, FTA, ETA, and finally a full QRA may be recommended.

This could happen because the organization believes that a proper evaluation of alternatives for more complex or severe risks requires more detailed tools. For example, LOPA is typically an order of magnitude risk analysis. Various alternatives may yield the same order of magnitude of risk (an enhanced LOPA may help in this regard) and the decision maker may want a better estimate of the reliability of various alternatives.

Background and guidance for establishing quantitative risk criteria is provided in *Guidelines for Developing Quantitative Safety Risk Criteria* (CCPS, 2009).

2.4.3 Risk Reduction Factor

The risk reduction factor (RRF) is a measure of how much a protective function or layer of protection reduces the frequency of a hazardous event. RRF is a means of communicating the performance of protective layers to people without sophisticated math and technical backgrounds (such as operators and businesspersons), without resorting to the use of negative exponents, which may be unfamiliar to them.

RRF is related to the average probability of failure on demand (PFD_{Avg}):

$$RRF = 1/ PFD_{Avg}$$

As a simple example, assume a facility or organization has target frequency for loss of containment of 1 in 10,000 (1×10^{-4}) per year. The current or baseline frequency of a scenario is estimated by PHA team or a frequency calculation, to be 1/10 (0.1) per year. To get to the target frequency, protective layers need a combined PFD of 1/1,000 (0.001) or an RRF of 1,000.

[Total PFD needed = 0.0001 (target frequency) / 0.1 (current frequency = 0.001]

RRF = 1/ 0.001 = 1,000

Two risk reduction alternatives are evaluated;

Alternative A has a PFD of 0.1 (RRF = 10)

Alternative B has a PFD of 0.01 (RRF = 100).

If the alternatives are independent, i.e., the performance of A is not affected by the performance of B, then:

Combined PFD = 0.1 x 0.01 = 0.001

Combined RRF is 1/0.001 = 1,000.

2.5 SUMMARY

The four steps in risk management are; risk identification, risk analysis, risk evaluation and risk treatment. The first three steps are covered in this chapter. A risk scenario is a sequence of events leading to a loss event. Scenario consequences and frequency can be estimated by methods ranging from engineering judgement to simple qualitative and quantitative tools to sophisticated computer models.

Risk evaluation requires an organization to establish risk criteria. The risk criteria need to be established at the highest levels in the organization, because a high consequence event affects the entire organization, not just the facility it occurs in. Likewise, every facility in the organization needs to know when to stop using resources to reduce risk at one place and use them elsewhere. Risk criteria can be qualitative or quantitative, and can apply to a single event, an entire facility, and the off-site population.

3

UNDERSTANDING PROCESS HAZARDS, CONSEQUENCES AND RISKS

3.1 PROCESS HAZARDS

Process Safety Knowledge is one of the 20 elements of CCPS' risk-based process safety management system (CCPS 2007). A thorough understanding the hazards of a process is fundamental to making good risk decisions. The following material is an overview of process hazards, and not meant to be a thorough discussion of the topics. The intent is to present information that relates to risk decisions. Resources for more material will be presented in each section.

3.1.1. Acute Toxicity

Toxicity is a measure of harm ranging from minor reversible adverse effects to major irreversible and life-threatening effects from caused by direct exposure to chemical substances. Acute toxicity refers to the effects of a single dose; chronic toxicity refers to repeated doses over time. Process safety risk management for process operations is usually concerned with acute toxicity, as risk scenarios are supposed to be one-time events/exposures. Chronic toxicity is a concern for repeated or ongoing exposures and is not the subject of the usual chemical process risk assessment.

Toxicity depends on the route of exposure, duration of exposure, and individual susceptibility. Any substance can be toxic given the right (or wrong) dose and exposure time. ("Poison is in everything, and no thing is without poison. The dosage makes it either a poison or a remedy" – Paracelsus, the father of toxicology, 1538.) Susceptibility cannot be factored into a risk analysis, so organizations need to consider the population exposed during a risk analysis. Personnel exposed to on-site risk are more likely to be healthy. The off-site population will have a range of susceptibility, so off-site impacts might be analyzed for the most susceptible population.

The most common routes of exposure in a chemical process operation are through the skin or the respiratory tract. Corrosive chemicals are an example of toxic materials that cause skin damage. Vapors and aerosolized materials are concerns for respiratory tract exposure. Asphyxiation is also a subset of toxicity.

Exposure Limits. For day-to-day operation the appropriate exposure index used in the United States is the Threshold Limit Value (TLV). TLVs are set by the American Conference of Governmental Industrial Hygienists (ACGIH). The TLV is meant for short term exposures:

- TLV-TWA (Time Weighted Average) for 8 hours a day, 40 hours a week
- TLV-STEL (Short Term Exposure Limit) for 15-minute exposures
- TVL-C (Ceiling) for maximum instantaneous exposure level.

Other countries have their own version of TLVs. Some are:

- Workplace exposure limit, UK
- Valeur limite d'exposition professionnelle (Occupational exposure limit value),VLEP, France
- Arbeitsplatzgrenzwert (Workplace Exposure limit), AWG, Germany

Some sources for workplace exposure limits are:

- American Conference of Governmental Industrial Hygienists (ACGIH, 2015),
- (http://www.acgih.org/forms/store/ProductFormPublic/2015-tlvs-and-beis)
- OSHA, Permissible Exposure Limits, (https://www.osha.gov/dsg/annotated-pels/index.html)

Exposure Limits for Risk Decisions. Most risk-based decisions deal with episodic events. For these events measures such as the Ambient Air Exposure Guidelines (AEGL) or Emergency Response Planning Guidelines (ERPGs) are frequently used for toxic end points. When a full Quantitative Risk Analysis (QRA) is done, an equation for the dose-response, known as a Probit equation is frequently used.

AEGLs are set by the National Advisory Committee for Acute Exposure Guideline Levels for Hazardous Substances (NAC/AEGL Committee), a US federal advisory committee. They are expressed as parts per million or milligrams per cubic meter (ppm or mg/m3) of a substance above which it is predicted that the general population, including susceptible individuals, could experience specified effects. These are:

- Level 1: Notable discomfort, irritation, or certain asymptomatic non-sensory effects. However, the effects are not disabling and are transient and reversible upon cessation of exposure.

- Level 2: Irreversible or other serious, long-lasting adverse health effects or an impaired ability to escape.
- Level 3: Life-threatening health effects or death.

ERPGs are set by the American Industrial Hygiene Association (AIHA). ERPGs estimate the concentration below which nearly all people will not experience the health effects defined by the ERPG level if they are exposed to the hazardous airborne chemical for 1 hour. A chemical may have up to three ERPG values, each of which corresponds to a specific tier of health effects. The three ERPG tiers are defined as follows:

- ERPG-3 is the maximum airborne concentration below which nearly all individuals could be exposed for up to 1 hour without experiencing or developing life-threatening health effects.
- ERPG-2 is the maximum airborne concentration below which nearly all individuals could be exposed for up to 1 hour without experiencing or developing irreversible or other serious health effects or symptoms which could impair an individual's ability to take protective action.
- ERPG-1 is the maximum airborne concentration below which nearly all individuals could be exposed for up to 1 hour without experiencing other than mild transient health effects or perceiving a clearly defined objectionable odor.

AEGLs and ERPGs for ammonia are provided in Table 3.1 as an example. Notice that the 60-minute AEGL and the ERPGs are not the same. This is not surprising since they are set by different organizations.

Some sources for these values are:

- Acute Exposure Guideline Levels for Airborne Chemicals (AEGL) http://www2.epa.gov/aegl
- Emergency Response Planning Guidelines™ (ERPG), American Indusial Hygienist Association, https://www.aiha.org/get-involved/AIHAGuidelineFoundation/EmergencyResponsePlanningGuidelines/Pages/default.aspx

Table 3.1 Ammonia AEGLs and ERPGs (ppm)

Level	1	2	3
AEGL 10-minute	30	220	2700
AEGL 30-minute	30	220	1600
AEGL 60-minute	30	160	1100
ERPG 60-minute	25	150	750
AEGL 4-hour	30	110	550
AEGL 8-hour	30	110	390

- TNO (The Netherlands Organization for applied research) Purple Book - CPR 18E - Guidelines for Quantitative Risk Assessment, (TNO 2007)
- Guidelines for Chemical Process Quantitative Risk Analysis (CCPS 1999)
- Lees' Loss Prevention in the Process Industries, 4th Ed., (Lee's 2014)

Probit equations. For some chemicals Probit equations may be available which will allow for more detailed risk assessments, particularly fully quantitative risk assessments, to be conducted. Probit calculations better assess the likelihood of impact on exposed individuals in that the calculations yield a probability of impact based on exposure time and exposure concentration. Further, in some situations it is also possible to assess potential impacts where the exposure concentration varies as a function of time.

Probit equations are developed by curve fitting dose-response data for a chemical to Equation 3.1.

$$Y = a + b \ln(C^n t_c) \qquad (3.1)$$

where:

Y = probit value
a, b, n = constants
C = concentration (ppm by volume)
t_c = exposure time (minutes)

The probit value, Y, relates to percentage of subjects affected as shown in Table 3.2. There are not many chemicals for which probit equations have been developed, and for some of them there are several different equations. Fixed concentration values such as AEGLs or ERPGs are available for many more chemicals.

Whether an organization decides to use fixed values or probit equations, or a combination of both in risk criteria and decisions, consultation with an industrial hygienist and/or a toxicologist is highly advisable.

Table 3.2 Probit Value vs Percent Affected

Probit Value	2.67	3.72	5.00	6.28
Percentage Affected	1	10	50	90

3.1.2 Flammability and Explosivity

Use of flammable and combustible materials creates the potential for fires and explosions. The fire triangle; fuel, oxidant, ignition source, is the classic way of considering the conditions needed for a fire.

To manage the risk of flammable gases and liquids one needs to know material properties such as the lower and upper flammability limits (LFL and UFL), the flash point (fp), autoignition temperature (AIT), limiting oxygen concentration (LOC), and minimum ignition energy (MIE) of the materials being handled. Many of these properties can be found on the Safety Data Sheets of the materials involved.

Inside equipment, processes may be run in a reduced oxidant atmosphere or at conditions where the fuel is below or above the LFL or UFL. Once a loss of containment event occurs, if a material is above its flash point there is a fire and explosion hazard because in air the oxidant is always present.

Control of ignition sources. Although controlling ignition sources is a good and necessary practice, is usually not considered a reliable risk reduction strategy by itself. In many investigations of fires and explosions, the ignition source is not identified with 100% certainty. Identification and elimination of all possible ignition sources is almost always not possible. This is why the first commandment of process safety is "Thou shalt always honor thy container" (Rosen 2015).

Once a flammable release occurs, it may or may not be ignited. If ignited immediate ignition occurs, there can be a fire or jet fire. If the ignition is delayed instead of immediate a pool fire, flash fire, or an explosion can occur. The degree of confinement of the flammable vapor cloud is an important parameter for determining if an explosion can occur, and some expert opinion should be sought in making that decision. For fires, the impacts of concern are flame impingement and thermal radiation from the fire. For explosions, the primary impact of concern is overpressure, however, other concerns include shrapnel and blast wave or impulse are effects. The event tree in Figure 3.1 is useful for thinking about the possible progression of a flammable release.

Combustible Dust. Combustible dusts are a special, and sometimes overlooked, case of the flammability and explosibility hazard. Finely divided organic or metal dusts can burn rapidly. A particle of 420 – 500 microns is frequently given as the size below which there is a concern for dust flash fires and explosions. This particle cut-off can be misleading; non-spherical particles such as flakes or fibers can be explosible even if one of the dimensions is greater than 500 microns Also, there is usually a distribution of particle sizes in a release, and while the average particle

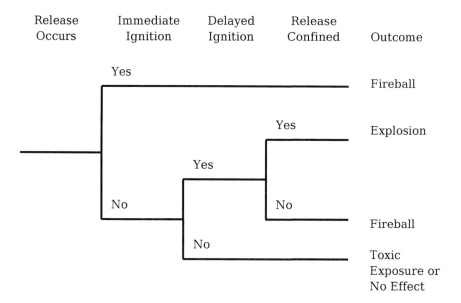

Figure 3.1. Event Tree for flammable release

size can be larger than 500 microns, a significant percentage of the dust can still be below that.

The requirements for combustion of a dust are the same as those for a vapor: fuel, an ignition source and an oxidant. However, dispersion and confinement of the dust are needed for an explosion or flash fire. Many of the parameters listed in section 3.1.2 have equivalent parameters with dusts.

As with vapors, any given dust cloud has an LOC and an MIE. Dust clouds have a Minimum Explosion Concentration (MEC) equivalent to the LFL of a vapor, and a Minimum Autoignition Temperature (MAIT), equivalent to the AIT of a vapor. Another key dust explosion property is the explosion severity index, K_{St}, which is needed to calculate the size of deflagration vents or of explosion suppression systems.

Unlike vapors, however, these properties are not intrinsic, but extrinsic properties, dependent on the particle size and size distribution of the dust in the cloud. In the standard test methods, the properties are usually measured on samples that have been screened to below 75 microns.

A key difference between vapors and dusts is the way confined explosions outside of the process equipment can occur. With vapors, loss of containment can immediately create a flammable vapor cloud which will persist until its concentration drops below the LFL. With a dust, leaks can accumulate on surfaces in a process rack or building. An initial event, such as an explosion in an equipment item, can disperse these deposits, which can then be ignited by the explosion. This creates a secondary explosion, as shown in Figure 3.2.

Secondary explosions can cause damage and injuries comparable to large vapor cloud explosions. In fact, all dust explosions that have caused multiplied fatalities and severe damage have been secondary explosions. For facilities handling combustible dust, a good housekeeping program is as important, if not more important, as a hot work permit program.

Without confinement, dust-based flash fires or fireballs can occur. Dust explosions can result when the fireball is confined, such as in an equipment item or in a building. Models for predicting the impacts are scarce for dust explosions. Therefore, simplified step logic is used for estimated consequence frequencies.

Boiling Liquid Expanding Vapor Explosion (BLEVE). A BLEVE is another type of explosion. It is defined as a type of rapid phase transition in which a liquid contained above its atmospheric boiling point is rapidly depressurized, causing a nearly instantaneous transition from liquid to vapor with a corresponding energy release. A BLEVE of flammable material is often accompanied by a large aerosol fireball, since an external fire impinging on the vapor space of a pressure vessel is a common cause. However, it is not necessary for the liquid to be flammable to have a BLEVE occur.

A common cause of a BLEVE is the failure of a vessel that loses mechanical strength when engulfed by a fire. In the liquid wetted wall of

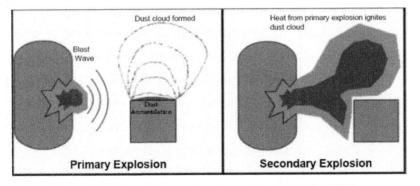

Figure 3.2. Secondary dust explosion (Courtesy of OSHA)

the vessel the heat is absorbed by the liquid. Where the wall is not liquid wetted, the temperature of the wall can rise to over 600°C, resulting in weakening of the walls. The metal of a pressure vessel can lose 60-80% of its strength, leading to vessel rupture.

The reader is directed to *Guidelines for Vapor Cloud Explosion, Pressure Vessel Bursts, BLEVE and Flash Fire Hazards* (CCPS, 2010) for more information about BLEVEs and BLEVE consequence models.

Pressure Vessel Burst (PVB). Pressure vessel burst (PVB) is a type of explosion that involves the bursting of a pressure vessel containing gas at an elevated pressure. The vessel does not necessarily have to be a pressure vessel as defined in the ASME Boiler and Pressure Vessel Code (BPVC). Any vessel that can build some pressure before bursting can cause a PVB. The ASME BPVC was designed to reduce the likelihood of PVBs.

Pressure Vessel Bursts can be caused by:

- Fabrication flaws
- Fatigue
- Corrosion or Stress corrosion cracking
- Cold temperature embrittlement
- Connection to a high-pressure supply
- Chemical reactions that generate pressure
- Combustion in the vessel
- Rapid phase transitions
- Overheating (e.g. due to external fire)
- Hydraulic rupture
- Human factors

When the vessel bursts, there will be a pressure wave; and fragments of the vessel will be dispersed. There are methods to predict the overpressure from a PVB. The reader is directed to *Guidelines for Vapor Cloud Explosion, Pressure Vessel Bursts, BLEVE and Flash Fire Hazards* (CCPS, 2010) for more information about PVB and the blast wave models.

3.1.3 Chemical Reactivity

Uncontrolled chemical reactions have the potential to result directly or indirectly cause serious harm to people, property or the environment. Many reactive chemistry incidents have occurred in operations where there was no intended chemical reaction, for example during storage, blending, and distillation. Chemical reactions can result in over pressurization of the containers or reactors, leading to:

- Equipment / Vessel ruptures
- Pressure waves
- Exposure to toxic or flammable materials

The key process safety concern in the design of reactors is runaway reactions. Runaway reactions occur when the heat generation rate froman exothermic reaction exceeds the rate at which heat can be removed, causing an uncontrolled rise in temperature. If the heat released by the reaction exceeds the cooling capacity, the reaction rate will accelerate (runaway) and may result in an excessive gas evolution or a vapor pressure increase that, in the absence of adequate overpressure relief protection, can rupture the reactor. If overpressure relief protection is adequate, then there will be loss of containment through the relief device.

During runaway reactions, the temperature can rise significantly, which may favor additional exothermic reactions. These may increase the rate of pressure rise, making it harder to safely relieve the pressure.

Guidance for safety in chemical reactor design can be found in the following sources:

- Guidelines for Engineering Design for Process Safety, 2nd Edition (CCPS 2012a)
- Guidelines for Process Safety in Batch Reaction Systems (CCPS, 1999a)

The US Chemical Safety Board (CSB) did an investigation of 167 reactive chemical incidents in the United Sates covering a 21-year period (CSB 2002). The CSB found that over 90% of the incidents involved hazards that were documented in publicly available literature. In response to the CSB study, the CCPS developed a 12-step preliminary screening method for chemical reactivity hazards and published it in *Essential Practices for Managing Chemical Reactivity Hazards* (CCPS 2003).

Other available tools and resources on chemical reactivity hazards include:

- Bretherick, Handbook of Reactive Chemical Hazards Vols. 1 and 2, Elsevier, 2007
- CCPS, Guidelines for Chemical Reactivity Evaluation and Application to Process Design, 1995
- CCPS, Essential Practices for Managing Chemical Reactivity Hazards (Johnson et al.) 2003
- CCPS Chemical Reactivity Evaluation Tool

- Compatibility charts, ASTM E2012-06, Standard Guide for Preparation of a Binary Compatibility Chart
- Chemical Reactivity Worksheet, Version 3.0, (http://response.restoration.noaa.gov/reactivityworksheet)
- ASTM Computer Program for Chemical Thermodynamic and Energy Release Evaluation (CHETAH®) version 10.0 (https://www.astm.org/BOOKSTORE/PUBS/chetah_support.htm)

3.1.4 Significant or Large Environmental Release Hazards

An environmental release is a loss of containment event which causes damage to the environment, such as contamination of soil or water or an air permit excursion. An environmental release may accompany the other hazards described above. Even without accompaniment by another hazard, significant environmental releases can cause millions of dollars of harm and will usually generate more bad publicity than a fire or explosion alone. Some extreme examples of environmental incidents are:

- Exxon Valdez (1989): Almost 11 million gallons of crude oil spilled from a grounded oil tanker into Puget Sound, AL. Exxon spent $2.1 billion on cleanup costs (Exxon Valdez Oil Spill Trustee Council).
- Sandoz Warehouse fire (1986): Water used to extinguish the fire in a chemical storage facility flowed into the Rhine resulting in a toxic chemical slick 40 km (25 miles) long, causing widespread destruction of aquatic life, which only began recovering more than a year following the incident.

3.1.5 Other Process Hazards

Examples of other process hazards that do not readily fit into the above sections are listed below.

High Stored-Energy Hazards. In high-energy rotating equipment (such as compressors, turbines, large pumps, centrifuges, etc.) issues such as improper alignment or imbalance can lead to equipment ruptures. Domino effects such as loss of containment and mixing of incompatible materials can also result from such failures.

Thermal Burn Hazards. Exposure to release of materials (e.g. superheated steam) that are extremely hot or cold can cause harm to people. At extremely high temperatures the strength of metals is lower than normal. Cold temperature embrittlement can cause equipment ruptures.

Asphyxiation Hazards. Low oxygen environments have caused many fatalities. A CSB study revealed 85 incidents of nitrogen asphyxiation resulting in 80 deaths and 50 injuries between 1992 and 2002. The CSB issued a safety bulletin and presentation on this hazard, *Hazards of Nitrogen Asphyxiation* (CSB 2003)

High- or Low-Pressure Hazards. Hazards of high pressure or vacuum that can harm people, and lead to equipment damage and releases. *Don't Become Another Victim of Vacuum* (Sanders 1993) provides several case histories of vacuum collapse of equipment.

Extreme Weather Hazards. Extreme weather is causing more losses over time. In 2017 Hurricane Harvey resulted in over 100 infrastructure failures and chemical releases. A method for quantifying and managing such risks is presented in *An integrated method for quantifying and managing extreme weather risks and liabilities for industrial infrastructure and operations* (Kytomaa et al. 2019).

3.2 RISK IDENTIFICATION

Not all hazard/risk identification tools are appropriate for scenario identification. Scenario based hazard identification methods focus on predictive methods to identify incident scenarios. Hazard identification tools that are good for predicting incident scenarios include Hazard and Operability (HAZOP), What If/Analysis, and Failure Mode and Effects Analysis (FMEA). These methods determine what can go wrong, and what safeguards are in place to prevent or mitigate the scenario. Table 3.3 shows a typical HAZOP worksheet with an example analysis of one cause of a loss of containment scenario. The causes, consequences, safeguards and recommendations columns facilitate a qualitative or quantitative analysis.

These tools can also be used to do risk analysis of the scenarios when supplemented with risk matrices, or they can be used to provide input for LOPAs. The book *Guidelines for Hazard Evaluation Procedures, 3rd Ed.* (CCPS, 2008), describes these tools and applies them to examples.

3.3 CONSEQUENCES AND IMPACTS

Depending on the type of hazards, the consequences can differ. Table 3.4 lists the hazards covered in Section 3.1 with their associated scenarios, methods that can be used to measure the scenario outcomes and potential impacts. Toxic, flammable, and environmental releases start with a loss

Table 3.3 Typical HAZOP review table format.

Deviation	Causes	Consequences	Safeguards	Recommenda-tions
More Flow	Inlet valve fails to close	Overflow of tank, toxic release	High level switch closes shutoff valve	

of containment event. A study team can qualitatively estimate the release size using judgement. The team can also use limited consequence modeling to determine an impact zone (e.g. the extent of the LFL and/or extent to a specified overpressure), or doing an assessment of the degree of confinement to cause an explosion, or apply dose-response relationships for toxic exposure, thermal radiation, and explosion overpressure. Some hazard scenarios can have more than one consequence. For example, A runaway reaction can result in a flammable and/or toxic release if an Emergency Relief System (ERS) functions successfully, or a vessel burst if the ERS Models exist for calculating the consequences and impacts for each column in Table 3.4.

3.4 FREQUENCY

For each hazard a frequency and likelihood can be estimated in a similar manner as consequences, i.e., using judgement, plant experience, failure rate data, and frequency calculation techniques such as LOPA or FTA, as described in section 2.3.3. Some hazards require estimating specific conditional modifiers such as probability of ignition for a flammable release. Other hazard scenarios can have more than one outcome. A runaway reaction of a given frequency can have two consequences as described above: a frequency for a flammable release from the Emergency Relief System (ERS), if an ignition source is found (the probability of ignition is needed), or a frequency of an overpressure event from a vessel burst if the ERS fails.

Table 3.4. Hazards and associated scenarios, measures and impacts

Hazard	Scenario	Measures	Impacts
Acute toxicity	Loss of containment	Size and composition of release Distance to specified exposure limit, ERPGs, AEGLs	Injuries, Fatalities
Flammability	Loss of containment and ignition Ignition inside an equipment item	Distance to specified thermal radiation levels Distance to specified overpressure levels	Injuries, Fatalities Building or equipment damage or destruction Injuries, Fatalities

Table 3.4. Hazards and associated scenarios, measures and impacts (continued)

Hazard	Scenario	Measures	Impacts
Chemical Reactivity	Runaway reactions with loss of containment or vessel rupture Inadvertent mixing with loss of containment or vessel rupture	Distance to specified overpressure levels Distance to specified thermal radiation levels Distance to exposure limits if an ERS is activated	Injuries, Fatalities Building or equipment damage or destruction Injuries, Fatalities
Pressure Vessel Burst	Elevated pressure inside a vessel	Distance to specified overpressure levels	Injuries, Fatalities Building or equipment damage or destruction Injuries, Fatalities
BLEVE	Fire engulfing vessel	Distance to specified thermal radiation levels Distance to specified overpressure levels	Injuries, Fatalities Building or equipment damage or destruction Injuries, Fatalities
Environmental Releases	Loss of containment	Size and composition of release	Soil and groundwater contamination, contamination of waterways.

Domino effect. Fire and explosion scenarios can initiate other scenarios. The fire or overpressure can cause damage to other vessels or buildings, creating more toxic and/or flammable releases, fires or explosions. The secondary dust explosion described in Section 3.1.2 is one example of these. An LPG terminal incident in Mexico, described below, is another example.

> **PEMEX Explosions and BLEVES**, Mexico City. On November 19, 1984, a Liquefied Petroleum Gas (LPG) release in a distribution terminal ignited and led to a series of explosions. The loss of containment was undetermined but could have come from overpressure of a pipeline or from overfilling a storage tank. The release of LPG continued for about 5-10 minutes when the gas cloud drifted to a flare stack and ignited.

> The degree of confinement in the horizontal storage tank enclosure was such that tanks were thrown off their supports and piping ruptured. Nine explosions and BLEVE's followed. Four small spheres were destroyed, and fragments scattered around the area, some as far as 350 meters (1150 ft.) away in public areas. The resulting fires led to the explosion of a series of the LPG storage tanks. About 600 people were killed, around 7,000 injured, 200,000 people were evacuated, and the terminal destroyed (*Incidents that Define Process Safety*, CCPS 2008a)

As an example of a simple semi-quantitative likelihood estimate, a decision team may need to estimate the likelihood of an explosion impact. The team could assume an order of magnitude estimate of the release frequency, and that delayed ignition and confinement needed for an explosion to always occur, i.e. have a probability of 1.0. The team can then use dispersion and blast models to calculate the distance to a given overpressure, such as 1 psig. For the impact the team can assume anyone in a building in the 1 psig blast zone becomes a fatality (1.0 psig can break windows and cause destruction to some structures. These are usually the cause of fatalities in an explosion.)   Although simple, this gives a conservative answer, in this case overestimates the likelihood of fatalities due to an explosion given a flammable release.

In this example, the decision makers then need to find alternative solutions that reduce the frequency of the release and ignition to below the organizations risk tolerance criteria.   If the team cannot find alternatives to accomplish this, it can then do more detailed analysis, perhaps adding an ignition probability estimate or a more detailed estimate of the release frequency, perhaps by FTA.

3.5 RISK

Severity and Likelihood can be qualitatively determined for any of the scenario outcomes described in Table 3.3. Once the consequence severity and likelihood are estimated risk and risk reduction requirements can be determined.

The frequency needs to match the consequence. A PHA team might classify a packing leak from a valve in high H_2S service as potential fatality (if someone is very close to the valve when the leak occurs, they could get a fatal dose of H_2S), and then base the frequency on how often that valve leaks ("that valve leaks all the time"). This overstates the risk. What is missing is either:

- the probability that someone is next to the valve when it leaks (a conditional modifier), or
- a frequency adjustment for the valve leaking enough to create a toxic zone that encompasses a normally occupied area.

Caution: When applying a qualitative methodology to the toxic release scenario above, teams may overstate the severity (no-one wants to go on record saying a fatality can't happen), but then intuitively compensate by understating the frequency (never seen anyone die from that). Maybe the reason that no-one ever died from it was not because the event (whether a toxic or flammable release) never happened, but because it was not high enough concentration to have a noticeable effect or reach a flammable concentration.

4

RISK DECISIONS AND STRATEGIES

4.1 OBJECTIVES AND ATTRIBUTES

4.1.1 Objectives

A risk decision implies there are competing alternatives that involve different levels of risk reduction and/or different costs. The decisions can be made in the context of a set risk tolerance criteria or as an assessment of relative risks of the alternatives. The risks most commonly thought in the chemical, petrochemical, refinery, and refinery upstream industry setting of are safety, environmental, and financial. Just as financial return targets for increasing financial risks are set at the corporate and/or business level, targets for tolerable safety and environmental risks also need to be set at the corporate level. The objective of a risk decision in the process industries is to reduce the safety, environmental, or financial risk of a process in a way that effectively uses the organization's resources and protects the implied license to operate.

4.1.2 Attributes

Degree of Difficulty. Some decisions may be relatively easy, such as installing a Safety Instrumented Function to prevent a process deviation that leads to a release of a toxic material that can cause one or more injuries or fatalities to people in the unit. Others may be complicated, such trying to decide whether to replace an atmospheric blowdown system with a flare or containment device or develop procedure to determine safe zones around the blowdown unit.

Organization Level. Risk decisions are made at all levels of an organization. Some examples are:

- A process deviation occurs, and an operator must decide whether to shut a process down or try to correct the problem while it is running.
- The maintenance and production departments need to decide whether to shut a plant down based on an inspection finding or run until the next planned shutdown. (See the following case study).

> **Chevron Richmond, WA refinery fire**. The failure of a 52-inch long carbon steel pipe released a flammable vapor cloud. The cloud ignited and engulfed 19 employees who, fortunately, were

able to escape. There had been several recommendations to replace the piping with more corrosion resistant piping, but Chevron never implemented them. (CSB 2014)

- A plant or engineering group identifies a unit that no longer complies with good engineering practice and must decide how and when to replace it. The decision may have to be made at a high level in the organization due to the costs and impairment of production. (See the following case study).

> **BP Texas City refinery explosion, 2005.** This is an example of a decision that needed to be made at a high level. A large flammable release through a blowdown system was caused by several operational and control system failures. The release resulted in 15 fatalities and major damage to the refinery. By the time of the incident, blowdown systems with atmospheric releases were no longer considered good engineering practice. The companies that owned the refinery, Amoco and later BP, intended to replace the blowdown drum but never did (CSB 2007 and Hopkins 2008). Replacement would have required capital investment and a production shutdown that likely would have had to be approved at a higher level than unit or plant management.

Decision Responsibility. Responsibilities for risk management and risk decisions need to be defined. Decisions for managing higher levels of risks need to go to higher levels in the organization. For example, consider a decision to postpone maintenance or replacement of an equipment item. If the equipment failure results in an event with local consequences, it could be made at the unit level. If the equipment failure could have plant wide consequences the decision may have to be made at the plant level, and where the consequences could have off-site impact, made at the corporate level. Clarity on risk ownership at facility level, site level and organization level and mechanism to escalate new and emerging risks based on the defined risk levels either due to management of changes, barrier impairment, external factors needs to be defined and understood.

Continuous Improvement. Risk decision strategies should comply with an organization's values/principles and be part of the organization's risk management processes. The risk management system should foster continuous improvement.

An example of driving continual improvement is BP's major accident risk (MAR) program. BP does not have an acceptable risk zone on its F-N curve (Considine and Hall 2009). Instead there is a single group reporting line (GRL) for each type operation, such as a refinery or an offshore oil well (see Section 10.2.3 – *Continual Improvement*). It drives continual improvement by requiring that the top 5 risks be addressed and reduced. Then, the next time the MAR is applied (5 years later) there should be a new set of the 5 top risks, and these must be addressed and reduced. Continuous improvement is expected even if the unit's risk is below the GRL. (Note, the MAR process was not applied to the Macondo Well. (CSB 2014a)

4.2 PROCESS LIFE CYCLE AND ALTERNATIVES.

The number and types of decision alternatives will vary with the different stages of a process lifecycle. The stages of a process life cycle are presented in *Guidelines for developing quantitative safety risk criteria* (CCPS, 2009a):

- Research
- Process development
- Detailed design and construction
- Operations, maintenance and modification
- Decommissioning

As listed in Section 2.4.2, The generally accepted hierarchy of risk reduction strategies is:

- Inherent – Eliminate the hazard by using non-hazardous materials and process conditions.
- Passive – Reduce risk by process and equipment design features that reduce frequency or consequences without the active functioning of any device.
- Active – Use controls, instrumented protection systems, or other devices (e.g., safety relief valves, suppression systems, automatic shutoff valves, automatic fire sprinkler systems) to reduce frequency or consequences.
- Administrative (procedural) – Use policies, operating procedures, inspection and maintenance procedures, hot work permits, and emergency response plans to reduce frequency or consequences.

The most risk reduction alternatives are available in the research and process development stages. Choices can include different raw materials, solvents, process conditions, and unit operations. Inherently safer options, e.g. less or non-flammable solvents, less toxic raw materials, reaction paths that bypass hazardous intermediates, fault tolerant processes/chemistries can be included.

By the detailed design stage, many choices are locked in. Alternatives at this stage could include plant location, equipment spacing and layout, equipment types, and the process control schemes. By the operations stage more of these alternatives are locked in and alternatives can include new procedures, new safety system interlocks, changes in raw materials, and equipment upgrades or changes. The more significant the change, the more likely it is to be a higher cost. A key point to remember is that at these stages, a Management of Change (MOC) review has to be done before implementing any risk decision.

Although application of inherently safer alternatives is less likely to be available in the operational phases (operations, maintenance and modification, decommissioning), they are not impossible. A potential risk decision can be to do a process modification that implements an inherently safer option. *Inherently Safer Chemical Strategies, A Life Cycle Approach,* 2nd Edition, (CCPS, 2009) is a source for information about inherently safer design.

4.3 THE DECISION PROCESS

One can find many decision-making processes in the literature, ranging from 4-step to 9-step processes. This book presents a modified version of the risk decision process presented in the earlier CCPS decision making book *Tools for Making Acute Risk Decisions* (CCPS 1995) with aspects of a process described in the book *Smart Choices* (Hammond, 1999). The process is:

- Define the problem
- Evaluate the baseline risk
- Identify the alternatives
- Screen the alternatives
- Make the decision

An astute reader should notice that if an organization is using the risk management system described in Section 2.1, the second step may already be done in the risk analysis and risk evaluation steps. There are situations, however, where this may not have been completed, as described in the next section.

4.3.1 Define the Problem.

The need to make a process risk decision can arise from a PHA, an incident investigation, from the realization that some current technology is no longer a good practice, or from observations by operations personnel, such as continued process deviations or control issues.

> **Morton International Patterson, NJ plant runaway reaction, 1998**. This runaway reaction at the Morton plant resulted in nine

> injuries and a shelter-in-place order for 10 square blocks. Operators noted in their logs in each of the 19 previous batches that exotherm had been difficult to control. An observant plant engineer reviewing the logs could have seen that there was a problem and halted production to initiate a risk decision process (CSB 2000).

During the problem definition, the risk decision maker should consider what the objectives of the problem are. Clearly, meeting the organization's risk criteria is one objective. Other objectives could be implementing the solution in a certain time frame, stay within a specified capital budget, minimize the testing and maintenance costs, and so on.

4.3.2 Evaluate the Baseline Risk

Risk Analysis was discussed in Section 2.3 and *Risk Evaluation* in Section 2.4. When a company comes to a risk decision through the risk management process as described in Chapter 2, the baseline risk may already be established. In a case like the runaway reaction example above, risk analysis and evaluation would have to be done at this stage.

Sometimes the evaluation of the baseline risk will identify if action is required or not, while later steps like the identification of alternatives are about finding the best action to take. In other words, the baseline risk assessment justifies the need to act, rather than justifying a specific action.

In other cases, e.g. response to an incident, a risk decision may need to be made even if previous reviews showed the risk was tolerable. In those cases, the previous hazard identification and risk analysis should be reviewed to see if there are lessons for improving the entire hazard identification and risk identification process.

4.3.3 Identify the Alternatives

A decision maker should want to identify a reasonable number of alternatives to ensure that they understand their options before choosing one to implement. Not identifying a possible alternative is missing an opportunity. If a company goes through *Step 2 – Evaluate the Baseline Risk*, some alternatives may present themselves during that stage.

A team approach, such as a brainstorming session, may be used to generate options. The team should be chosen in the same way one would choose a PHA team. Representatives from operations, and people with technical skills appropriate to the problem, such as chemistry, engineering design, process control, etc. should be included as needed. An individual familiar with the risk assessment technique being used should also be part of the team. See *Guidelines for Hazard Evaluation*

Procedures, 3rd Edition for more information about choosing a team (CCPS, 2008).

As options are generated and evaluated, a review of the original problem definition and objectives may be warranted.

4.3.4 Screen the Alternatives

Much of the rest of this book is about screening the alternatives. The risk analysis and evaluation steps have to be applied to the alternatives.

Stepwise Options and Alternatives. When determining if a solution meets the organization's risk criteria, a good practice for risk analysis and evaluations is to start with a simple tool such as a risk matrix, and advance through more sophisticated consequence and frequency analysis, and perhaps progressing to a full QRA, only if needed (see Section 2.3).

The objectives and outcomes (Section 4.4) and tradeoffs (Section 4.5) of each alternative need to be evaluated (Hammond 1999). Some alternatives can hopefully be dismissed as the progression through these steps proceeds.

4.3.5 Make the Decision

When a decision is made, the uncertainties, risk tolerance, and the effect of linked decisions should be identified and examined. These topics are covered in Sections 4.5, 4.6 and 4.7.

4.4 OBJECTIVES AND OUTCOMES

The outcomes of each alternative describe how well it meets the objectives laid out in the problem definition stage. Meeting the organizations risk criteria is only a minimum objective. Some typical objectives can be:

- Amount of risk reduction
- Reliability of the alternative
- Capital cost
- Operating cost (e.g. inspection, testing, and preventive maintenance)
- Event impact (e.g., fatalities, property damage, cost of cleanup, community impact)
- Likelihood of success
- Timeline to implement

In *Smart Choices*, (Hammond, 1999) the author recommends creating an objectives table. For the objectives above, this would look like Table

4.1. The numbers in Table 4.1 are made up for illustrative purposes only. By inspection, one can deduce that Alternative A is an administrative option, perhaps operator response to an alarm, and Alternative D is likely a design modification, perhaps an inherently safer design that essentially eliminates the event.

As alternatives are being screened, some may be eliminated because they cannot meet one or more of the objectives. After developing an Objectives and Outcomes table it may be possible to make a decision by inspecting the table.

4.5 TRADEOFFS

After a consequence table has been made, if there is more than one alternative left, then tradeoffs between the alternatives are evaluated. This is because it is very probable that no one alternative will be the best for all objectives. How the tradeoffs between objectives are made will depend on an organization's culture, operating philosophy, size, and so on.

In the Table 4.1 case, for example, a global, multi-billion-dollar organization with high public visibility might choose Alternative D because the elimination of community impact outweighs the high capital cost. Another organization may eliminate Alternative D because the business cannot support the investment. This organization must make a tradeoff between the consequences of Alternatives A, B and C. In these cases, the tradeoffs are between risk reduction, costs, and potential damage and community impact.

One way to handle these tradeoffs is to use a scoring method. A very simple ranking is presented in Table 4.2. Each option is ranked on how well it meets the objective, the best being 1, the worst 4. From Table 4.2, one would think that Alternative D is clearly the best. This simple ranking ignores the relative importance capital cost, however. As noted before, a business may not be able to support the high investment.

Table 4.1. Example Objectives and Outcomes Table

Alternative	A	B	C	D
Risk Reduction	10X	100X	100X	1000X
Reliability	90%	99%	99%	99.9%
Capital Cost	$0	$100,000	$10,000	$1,000,000
Operating Cost	$1,000	$5,000	$10,000	$0
Damage Impact	$500,000	$10,000	$50,000	None
Community Impact	High	Low	Medium	None

Table 4.2. Simple Ranking for Table 4.2 Alternatives

Alternative	A	B	C	D
Risk Reduction	3	2	2	1
Reliability	3	2	2	2
Capital Cost	1	3	2	4
Operating Cost	2	4	3	1
Damage Impact	4	2	3	1
Community Impact	3	2	2	1

Other weighted scoring techniques are described in *Tools for Making Acute Risk Decisions* (CCPS 1995) and include the advanced hierarchy process (AHP), multi-attribute utility analysis (MUA), and simplified multi-attribute utility analysis technique (SMART).

A weighted scoring technique can be used to deal with the relative importance of each objective. One such technique is the Kepner-Tregoe method (Kepner and Tregoe, 1981). In this method, the objectives are first divided into Musts and Wants. Meeting or exceeding the risk tolerance criteria is always a must. Other examples of Musts might be: the capital cost must be less than $500,000, and the risk reduction must be least a factor of 100. Any alternative not meeting these are eliminated. In the example from Table 4.1, Alternatives A and D are eliminated on these criteria. The remaining objectives (Wants) are weighted or ranked from 1 to 10, with 10 being the most desirable. Each alternative is then given a score of 1 to 10 for how well it meets each Want.

For each alternative, the weighting of each objective is multiplied by its score, and the results are added to give a total weighted score. The option with the highest score is the best.

For the example in Table 4.1, assume that the company is very risk averse and prioritizes risk reduction over other factors. The weights for each want are assigned as:

- Community Impact   10
- Damage Impact   8
- Capital Cost   6
- Operating Cost   4

Since the reliability of alternatives B and C is the same, this objective did not have to be included in the weighting. Alternatives B and C are then scored on how well they meet the objectives. A final scoring and resulting weighted score are shown in Table 4.3. With the scoring in Table 4.3, Alternative B is better.

Table 4.3. Weighted Scores for Example from Table 4.1

Objective	Weight	Alternative B		Alternative C	
		Score	Weighted Score	Score	Weighted Score
Community Impact	10	8	80	6	60
Damage Impact	8	9	72	7	56
Capital Cost	6	4	24	9	56
Operating Cost	4	9	36	8	32
Weighted Score			212		204

In a weighted scoring method, the real decisions are the weights for the objectives and the scores given the alternatives. When dealing with objectives such as Community Impact, or Capital Cost, one individual should not be making all the decisions. The decisions should be made by a team that will consist of representatives whose viewpoints will include the appropriate perspectives from the corporate, business, and operations level of the organization.

4.6 UNCERTAINTY

There will always be uncertainty in risk decisions. To deal with uncertainty, one needs to ask four questions (Hammond 1999):

1. "What are the key uncertainties?
2. What are the possible outcomes of these uncertainties?
3. What are the chances of the occurrence of each possible outcome?
4. What are the consequences of each outcome?"

There are several sources of uncertainty in the information used by process risk management and decision tools and hence the decision itself.

Incomplete definition of the risk problem. This is a potentially large source of error. If the risk problem came from a PHA, and the PHA team does not identify some hazard scenarios, they can go unprotected and not even be considered in a risk decision. To improve the potential to identify as many relevant scenarios as possible, the PHA team should have members with experience in operations, maintenance and engineering. Experts with knowledge in chemistry, instrumentation and control, and other fields should at least be available as needed, if not on the team itself. The book *Guidelines for Hazard Evaluation Procedures, 3rd Ed.* (CCPS, 2008) provides more information about selecting a PHA team.

Missed scenarios may be found when PHAs are revalidated or by investigating and learning from near misses or incidents. If a new scenario is identified in a PHA that has not been considered before, then the scenario should be reviewed for how it affects any existing risk analysis to see if that analysis needs to be updated.

A good practice for near misses is to investigate and ask what would have happened if a safeguard failed. If the resulting consequence can affect an existing risk analysis, that risk analysis should be updated.

Inadequate process knowledge. Not understanding possible hazards in a process is also a potentially large source of error. This feeds into the problem of incomplete definition of a risk decision. Reactive hazards are good examples of this (see the T2 Laboratories example). As an example, if an organization does not know or identify a potential interaction between a process chemical and a heat transfer fluid, or an absorbent, then the need for a risk decision will not be identified at all. Again, the best way to improve the chances of not missing possible hazards is to have a team with experience in several fields of expertise.

> In 2007, a chemical manufacturer, T2 Laboratories, Inc. (T2), was making a batch of methylcyclopentadienyl manganese tricarbonyl (MCMT), when an explosion occurred, killing four people and injuring 32. Investigators from the Chemical Safety Board (CSB) concluded there was a loss of cooling that led to a runaway reaction. Operators did call for technical support from the company owners. The explosion occurred minutes after their arrival. The CSB concluded that "T2 did not recognize the runaway reaction hazard associated with the MCMT it was producing." The failure to recognize the hazard led to a cooling system design that was not reliable and a reactor design that did not have an emergency relief system that could safely relieve the pressure. (CSB, 2009)

Margin of Error - Consequence analysis. As with frequency analysis, a qualitative consequence analysis by a PHA team can be wrong. Uncertainty may be driven by not wanting to understate the outcome (no PHA team wants to hear that they failed to recognize the severity of an outcome especially after an incident found it the hard way), or it may overstate severities. Their familiarity with the process hazards may lead them to underestimate the consequence because it was not that bad when it happened in the past (which doesn't mean it could not have been worse.)

George E.P. Box, a highly regarded statistician, said "essentially, all models are wrong, but some are useful." This holds true for consequence analysis models. These models have many assumptions built into them which the modeler may or may not be able to adjust. For example, most dispersion models are essentially open field models; the effect of buildings or obstacles on the dispersion is not considered. If necessary, there are Computation Fluid Dynamic (CFD) models that can better do this.

As a result of the assumptions in them, the distance a dispersion model calculates to a concentration of interest, such as the LFL of a flammable material, is only an estimate. This should be considered when making risk decisions or when establishing risk criteria. In the case of a flammable release, an organization can decide to use the distance to ½ of the LFL as a criterion.

An example toxic consequence severity criterion could be that concentrations of a material above a specified level, such as the ERPG-2[1], should not go past the plant boundaries. If a model shows that the ERPG-2 is just inside the plant boundary, an organization should consider the potential margin of error before making a risk decision. For example, if there is empty land on the other side of the boundary, this may be tolerable, but if there are occupied buildings or a frequently traveled road, this may not be tolerable. When looking at the potential for exposure to toxic materials, exposure time may also need to be assessed. For example, will offsite populations be exposed to a prolonged event or will the exposure be highly transient in nature?

Margin of Error - Likelihood calculations. Initiating event frequency estimates can be too high or too low by orders of magnitude. If the scenario consequence is high, e.g., a fatality, loss of a unit, or severe impact to the public, qualitative estimates of frequency and probability should be challenged.

LOPA is usually done as an order of magnitude calculation (unless the organization uses an enhanced LOPA), so the initiating event frequency is sometimes rounded off to a higher value to start with. Failure probabilities of the safeguards are also rounded off (usually 0.1, 0.01, etc.). Therefore, different alternatives can have the same calculated likelihood. If it becomes important to account for differences in initiating event frequency or the reliability of layers of protection, perhaps because the alternatives have high cost, then better estimates of the likelihood from an FTA, ETA, or QRA may be needed.

FTA, ETA, or QRA frequency calculations have their own sources of error, including the basic event data. Initiating event frequency data for

---

[1] ERPG-2 is the maximum airborne concentration below which it is believed nearly all individuals could be exposed for up to one hour without experiencing or developing irreversible or other serious health effects or symptoms that could impair an individual's ability to take protective action.

an FTA or QRA should reflect plant conditions whenever possible. Care needs to be taken, especially when only minimal data is available, since the frequency data may not be statically sound and may reflect underlying system noise rather than providing a reliable source of frequency data.

Many organizations do not collect failure rate data systematically. Therefore, the fallback becomes using generic data from various sources. Whatever the data source, the basic data used in calculating frequency of alternatives should be equivalent. That way the relative differences between alternatives should be meaningful. Sources of failure rate data include:

- UK HSE HCRD dataset (https://www.hse.gov.uk/hcr3/)
- IEEE Gold Book Standard (IEEE 2007)
- IEEE Standard 500 (IEEE 1984)
- TNO Dutch Purple Book (TNO 2005)
- OREDA Off-shore Reliability Data (OREDA 2015, 6th Edition)
- Guidelines for Process Equipment Reliability Data, with Data Tables (CCPS, 1989)

Capital and operating costs. All cost estimates have a margin of error. The margin of error on initial capital cost estimates can be high enough that the relative costs of two alternatives may be essentially the same. The amount of error will decrease as more detailed engineering design work is done. Decision makers should discuss the cost estimates with those doing them to understand the uncertainties in those estimates.

Residual risk. There is no such thing as zero risk, so even after implementing the risk mitigation recommendations, risk will remain. Companies need to assess if this residual risk is tolerable and if further risk reduction may still be required. That is why the concept of continued risk reduction is included in ALARP. If reasonable risk reduction methods can be identified, they should be implemented. However, if a plant has identified scenarios in other units that have a higher risk after implementing a decision in a given unit, for example by qualitative using a risk matrix, reducing those risks first is a reasonable risk decision.

4.7 RISK TOLERANCE

In discussing the tolerability of risk, the UK HSE has written:

"... 'tolerable' does not mean 'acceptable.' It refers instead to a willingness by society as a whole to live with a risk so as to secure certain benefits in the confidence that the risk is one that is worth taking and that it is being properly controlled. However, it does not imply that ... everyone would agree without reservation to take this risk or have it imposed on them." [HSE 2001]

This caveat should be considered when making a risk decision. Even if the organization has established risk tolerance criteria, there are some occasions when the final risk decision should be reviewed against other parameters.

Previous events at the site's location, even if not caused by your organization, should be considered. Consider the incident below.

> **Freedom Industries, WV environmental release (2014)**. Approximately 7,500 gallons of 4-Methylcyclohexanemethanol (MCHM) leaked from a storage tank into the ground, and eventually into the Elk River in West Virginia, about 1 mile upstream of the intake to the local water treatment system. The resulting contamination left about 300,000 residents without potable water.

Given the Freedom Industries incident, if an organization has to make a decision at a site near a river in West Virginia or any river for that matter, it may choose to apply a stricter criterion for a release, as a similar incident would certainly provoke public outrage and potential backlash.

Sensitive public areas near a site may warrant further consideration of the risk tolerance. If the adjacent area is another chemical, petrochemical, or industrial site, emergency response agreements can be considered in the final decision. If the adjacent area is a residential neighborhood, school, hospital, etc., the organization may want to consider a stricter risk tolerance criterion and/or take no credit for emergency response.

4.8 LINKED DECISIONS

A linked decision is one in which the decision will have an effect on a decision in the future. Examples in the process industries could be:

- Choice of chemistry and solvents
- Choice of unit operations
- Choice of location and layout
- Choice of risk reduction options.

Identifying the future decisions that may be linked to the initial decision is the first step in dealing with linked decisions. For example, an organization's research division may be working on process improvement involving a change in the chemistry and/or process solvent. One risk reduction alternative for the existing process may preclude that change while another enables it. One way of handling this dilemma is to include enabling the future change as one of the objectives to include when making tradeoffs. Another can be to postpone the decision until the research group has had more time to complete its research.

Key to identifying linked decisions in a large organization is communication between the various groups that have an interest in the decision; research, engineering, operations, business and corporate.

4.9 DECISION TREES

Decision trees are a tool that can help a decision maker deal with uncertainties and linked decisions. A decision tree resembles an event tree; however, a node can have more than two branches. Figure 4.1 is a decision tree that looks at possible branches for the damage estimates for Alternatives B and C in Table 3.1. For this example, it is assumed the damage estimate of each alternative is ± 50% due to use of a very simple consequence/impact model.

In Figure 4.1 it can be seen, that the worst case for Alternative B is lower than the best case for Alternative C, and that Alternative C can be much worse. With these assumptions, the differences in damage impact between the two alternatives are clear. In some cases, it may be possible there will be overlap between some of the potential values for each alternative. When that happens, the decision team will want to have some idea of the probabilities of each final case, A through F. The decision team needs an estimate of the likelihood of each branch from the specialists who did the frequency, consequence or cost calculations for each alternative.

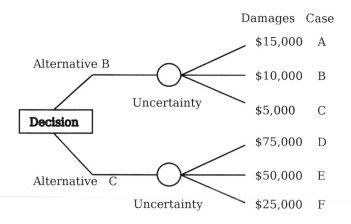

Figure 4.1. Damage Impact Decision Tree

In a similar manner, a risk decision team can evaluate linked decisions by extending the decision tree. The linked decision can be shown at the end of the consequence branches of the first decision with the resulting consequences of that decision shown as well. Even if probabilities of every branch cannot be determined the visual depiction of the decision alternatives and their consequences can be helpful to a decision team.

Evaluating baseline risk was discussed in Chapter 2. Identification of alternatives needs to be done with an open mind. A problem in many risk decisions is that organizations can may default to the first alternative identified.

Alternatives are screened by evaluating their consequences and deciding how to make tradeoffs between competing consequences. Decision teams need to consider the uncertainties in the consequence estimation and to identify future decisions that may be linked to the one under consideration.

Tools such as weighted scoring methods and decision trees are useful for examining tradeoffs between consequences, uncertainties and linked decisions.

5

DECISION MAKING

5.1 DEFINING THE DECISION PROBLEM

In process safety the decision problems can arise from many sources, including; recovery from a process deviation, a process change, a hazard assessment finding (e.g. What-If Checklist review, HAZOP, facility siting, or MOC review), an audit or inspection finding, incident investigation findings (e.g. from a Root Cause Analysis (RCA) investigation), alternatives for a process design, or the need to upgrade a facility after an acquisition or merger.

In many decision problems, process safety considerations are mixed with other considerations. Examples include, but are not limited to, impact on operating and capital cost, maintaining equipment reliability, increases in process complexity (due to additional safety devices), consequences of potential incidents, and time pressures. These form the list of objectives as described in Section 4.4 – *Decisions Objectives and Outcomes*. Therefore, risk decisions will likely involve a balance of safety, environmental, and business risk priorities. The more aspects involved, the more complex the problem and the more sophisticated the decision tools will become.

5.1.1 Types of Decisions

For this book, decision problems are grouped into three types, (1) quick and/or simple, (2) intermediate, and (3) complex decisions. The dividing line between them is not always a sharp line, these three types should be looked at as a continuum of increasing complexity.

Quick and/or Simple Decisions. Quick decisions are characterized by the need for quick action. A decision to shut down a process is a yes/no decision, and seemingly simple, but failure to do so quickly can result in a serious consequence. There can be a few alternatives for a quick decision. An example is response to a process deviation (do I apply cooling, stop feeds, or add a shortstop agent to the batch?). Consultation with a few other people may be possible.

Simple decisions are characterized by few, usually well understood, alternatives and have no time pressure. Simple does not necessarily mean low risk. Consultation with others is possible. Some design decisions (what type of pressure relief device should I use?) can be simple decisions.

Complex Decisions. These are characterized by several alternatives, frequently with high costs associated with them, and several competing objectives and outcomes. Deciding between the solutions requires a lot of information and time to generate the data and make sometimes multiple stepwise decisions. Complex decisions usually require quantification of consequences or frequencies or both (i.e. QRA). These are group decisions, and may need input from many groups, such as, business and corporate management, plant departments (production, engineering, safety, maintenance), and corporate staff (engineering and research).

Examples of complex decisions can be a major modification or upgrade of an existing process, relocating a process (such as tolling), closing a production unit, or decisions about processes that pose risk to many people on or offsite, and/or significant business risk.

Intermediate Decisions. The intermediate type represents a continuum of decisions between the simple and complex types. There is some gray area between intermediate and complex decisions, so the dividing line is a little harder to define. They have several characteristics that can vary.

Time to implement. Action on the order of minutes is not necessary, but anything from a few hours to a few months may be needed.

Level of Approval. For complex decisions it was noted that several groups may be involved in the decision. This implies that the decision maker is probably high up in the organization hierarchy, perhaps a business manager or higher. Intermediate decisions can be made at a lower level, for example within an operating unit or plant (e.g. area or plant manager) or technical unit (engineering or research management). The level of approval is frequently linked to the operating expense or capital cost involved.

Number of people. The size of the group involved is directly related to the level of approval. An intermediate decision can be made individually or by a few people. The magnitude of the decision is probably such that it can be made within a limited group, such as an operating unit, a manufacturing plant, staff or corporate engineering, or research.

Qualitative vs. Quantitative. The full range of qualitative and quantitative risk tools can be involved. Risk matrices or LOPAs can be used. Some consequence modeling may be needed.

Examples of intermediate decisions include:

- Implementation of a hazard assessment, MOC, audit or incident investigation recommendation (e.g. installing an interlock or correcting/updating an operating procedure)

- Modifying a unit operation
- Changes in the staffing (e.g., manpower, length of shifts)
- Installing a new equipment item
- New plant and equipment layout
- Upgrading a control system

Table 5.1 summarizes characteristics of the three types of risk decisions.

5.2 SELECTING A DECISION TOOL

5.2.1 Progression of Risk Analysis Tools

When selecting risk evaluation tools, preference should be given to the simplest tool possible. The preferred progression is:

- Qualitative hazard evaluation (e.g. What-If, HAZOP, FMEA) using engineering judgement or qualitative risk matrix
- Semi-Quantitative Analysis:
    o Hazard evaluation techniques (e.g. What If, HAZOP, FMEA combined with a Risk Matrix (can be qualitative or semi – quantitative)
    o LOPA (in the US) or Risk Graph (in Europe)

Table 5.1 Decision Characteristics

Quick	Simple	Intermediate	Complex
Single scenario	Few alternatives	Limited number of alternatives	Multiple scenarios
Limited time available	Low cost alternatives	Low/medium cost alternatives	Several alternatives
Few (perhaps only 2) alternatives	Well understood scenarios	Several objectives and outcomes	High cost alternatives
	Few or simple objectives and outcomes	Obvious/simple trade-offs between objectives	High consequence alternatives
	Few or no trade-offs	Possible linkage or domino effect	Several objectives and outcomes
	No linkage (domino effects)	Basic decision tools enough	Complex trade-offs between objectives
	Engineering or experienced based judgment (or team voting) enough		Domino effects
			Linkage to future decisions
			Complex tools needed to evaluate alternatives

- Quantitative Methods
  - o   Consequence Analysis
  - o   Likelihood quantification (Fault Tree Analysis, Event Tree Analysis)
  - o   Cause-Consequence Analysis
  - o   Quantitative Risk Analysis (QRA)

Inadequate forethought can lead to incorrect characterization, putting a higher-level decision into a lower category, with serious possible consequences.

Bridges and Dowell (2016) state that qualitative analysis is appropriate for 95% of risk decisions, semi-quantitative for 5%, and quantitative analysis for 0.01%. The qualitative evaluation must be done by a competent team for their assertion to be considered valid and does not apply to processes where consequence modeling is required (for example facilities in the US covered by EPA's RMP rule or offshore facilities in Norway). Process safety professionals may disagree about the exact percentages Bridges and Dowell provide. Most would agree, however, with the basic idea; most risk decisions, perhaps 80-90%, can be made using a qualitative hazard analysis, with perhaps 5-18% needing a semi-quantitative method, and a few (2-5%) needing a quantitative method.

A decision maker or team may find they will conduct an iterative process of analysis with increasingly complex tools.

5.2.2 Factors in Decision Tool Selection

Factors that need to be considered in selecting the decision tool include:

- The amount of time available to reach the decision
- The level of approval needed for the decision
- The resources (manpower) needed for the decision
- The magnitude of potential scenario consequences and the risk receptors
  - o   On-site only or on-site and off-site
  - o   Safety, environmental, business, reputation, or a combination of these
- The complexity of the decision
  - o   Number of alternatives
  - o   Types of decision outcomes (Section 4.4)
  - o   Linked decisions,
  - o   Implementation requirements (resources needed to implement the selected alternative)
  - o   Future impact (Is this a short-term issue or will this set direction for the company?)
- Impact of local regulations

Most of these factors are also attributes of the type of the decision problem, simple, intermediate or complex.

Quick and/or Simple Decisions. As already stated, deciding how to address a process deviation is an example of a quick decision. Operators and front-line supervisors encounter process deviations such as leaks, instrument and equipment malfunctions, or processes operating outside their normal operating ranges during their work. The time scale to resolve these can be a matter of minutes. Cost, in the form of lost production, may loom large in an operator's mind, while the magnitude of the consequences may not be known or understood. For some quick, simple decisions a troubleshooting or Consequence of Deviation guide or checklist, developed ahead of time should be used. Application of the appropriate risk analysis and decision-making tools described in Chapter 2 can be used to create such a guide. In a sense the troubleshooting guide is the outcome of an intermediate decision. Troubleshooting process deviations requires more training of both operators and the production engineers. Both drills and table-top exercises are valuable means of determining if decisions require updates to training. The Esso Longford explosion provides an example of how inadequate training can lead to poor decisions.

> **Esso Longford Gas Plant Explosion (AUS, 1998).** A pump trip caused loss of hot lean oil flow to a heat exchanger caused a drop in temperature in the exchanger (to -48 °C (-54 °F)), causing embrittlement of its shell. Reintroduction of hot oil to the exchanger led to its rupture and a flammable release which resulted in an explosion. Loss of the lean oil was a critical event, but operators were not aware of this because a hazard assessment of the plant had never been done. The existence of a troubleshooting guide created from a hazard assessment could have prevented this. *Incidents That Define Process Safety* (CCPS 2008a)

Following Recognized and Generally Accepted Good Engineering Practice (RAGAGEP) or internal design standards (which may be based on RAGAGEP) to make design decisions is an examples of a quick, simple decision for process engineers (analogous to the trouble shooting guide for operators). However, it is not recommended to use pieces of standards or codes without regard to their full applicability. Relying solely on conformance to standards and codes does not eliminate the need for a risk decision, many codes represent minimal requirements only and may not meet an organization's risk tolerance criteria.

> *Recognized and Generally Accepted Good Engineering Practices* are the basis for engineering, operation, or maintenance activities and are themselves based on established codes, standards, published technical reports, or recommended practices (RP) or similar documents. RAGAGEP details generally approved ways to perform specific engineering, inspection or mechanical integrity

activities, such as fabricating a vessel, inspecting a storage tank, or servicing a relief valve. (CCPS definition)

Many design decisions can be aided by using RAGAGEP for the technology being used by the plant or business as a tool. By identifying and documenting appropriate RAGAGEPs, an organization can standardize many simple design decisions. An organization may apply any of the risk analysis tools and decision aids described in Chapters 2 and 4 to develop its RAGAGEPs.

Intermediate Decisions. Section 5.1.1 listed some types of intermediate decisions.

For those listed items the complexity of the problem and the magnitude of the scenario consequences can range from minor to catastrophic. All of the risk tools described in Chapter 2 can be used. The factors listed at the beginning of this section will feed into the selection of the risk tool needed.

When qualitative risk judgement is sufficient, risk matrices or engineering judgement are the tools that can be used.

If an organization uses quantitative risk measures, the intermediate tools available are quantitative risk matrices, consequence modeling, FMEA, LOPA, risk graph, FTA and/or ETA. If the last few tools are needed, the decision may move into the complex category. The selection of the tool depends on the size and complexity of the decision. For example:

- If the decision depends on the consequence of the scenario alone, consequence modeling may provide enough input for the decision makers.
- If the consequence is known, or does not need to be quantified in detail, frequency analysis tools such as a quantitative risk matrix, LOPA, or FTA may be sufficient.
- If a scenario has a few possible consequences and the likelihood of each is needed an ETA is appropriate (or a LOPA can be done for each consequence).

Remember, there is a continuum of decision types and tools. In some cases, a progression may occur where the simplest risk tool is applied first and then risk tools of increasing complexity are applied until such point that a clear decision can be supported.

With intermediate decisions the individual or team can progress through all of the decision tools and steps described in Chapters 2 and 4.

Complex Decisions. As mentioned earlier, a complex decision may require a full QRA. An organization may reach this conclusion after trying to make a risk decision by progressing through the various risk tools as described in the intermediate decision section. An organization may also start out knowing a decision is a complex one and opt to go directly to a QRA. It is very likely that a weighted scoring tool will be needed to assess the outcomes and balance the priorities of several alternatives with respect to safety, environmental, and business risk. An example could be a large-scale LNG liquefaction facility where a preliminary QRA might be done at early design time to establish basic process safety, with a detailed QRA done later in design to verify those decisions and to identify any other issue. Use of a company that specializes in doing QRAs brings an unbiased assessment and can provide more credibility to the decision.

5.3 ASSEMBLING THE APPROPRIATE ASSESSMENT RESOURCES

The appropriate resources, people, tools and time, need to be made available for the decision.

5.3.1 Team Members

The size and make-up of the decision team are a function of the type of decision. Like a PHA team (CCPS 2008), the members selected are a function of skills needed to analyze the problem and define solutions, and the level of responsibility required to authorize the decision team's recommendations.

Stakeholders, people or organizations who can be affected by the risk decision, should be represented. For example, production and plant stakeholders such as engineering, operations, maintenance, safety, health, and environmental managers, and the business may need to be involved. Also, the local community may be affected and hence should be represented in some way. This may be through a corporate or site representative who has the best interest of the company and community in mind. Or, this could include involvement of the local emergency planning committee (LEPC) or community advisory panel (CAP). Technical experts, either internal or external, may also be needed depending on the complexity of any analysis needed. The team composition should be appropriate to the level of risk and the complexity of the potential solutions and include individuals familiar with the tools and techniques required to complete the risk assessment.

Quick or Simple decisions, by their nature, involve only one to three people, for example a plant engineer and a reliability engineer could decide if an alternative pump for replacement is acceptable. The approval level is local, perhaps the process engineer or a unit manager. A process engineer and control engineer can decide what kind of instrument is appropriate for a control loop. The approval could be from a senior process engineer or engineering manager. If more than three people are needed to make the decision that is a sign the decision is more complicated than originally thought.

Intermediate decisions involve more detailed analysis. A small team of people is needed. In a plant, a decision on how to proceed with batch that has a stalled reaction could involve the plant engineer, a process chemist, the plant HSE manager, and the unit manager. The unit manager could have the final approval. An example a decision involving a design option can involve a process and project engineer, a plant engineer, and a process safety specialist.

Complex decisions will need a robust team. The risk analysis tools can require a specialist, especially if a quantitative method (FTA, ETA, or QRA) is needed. If this person is an outside contractor, then an internal process safety representative is needed to ensure the company's risk criteria and methods are followed. Options with different capital and operating costs are likely to be involved, so a team member familiar with capital cost estimates and cash flow analysis is needed. Different design options imply that a process engineer is needed. If the decision involves a plant addition or expansion, a plant engineering and manufacturing representative is needed.

Some items to consider in creating the team are:

- *Capital or Expense Cost.* How high up is the approval level for the capital involved? That person either should be on the team or have a representative.
- *Decision tool.* What is the decision tool selected, is it simple or complex? An expert with that tool may be needed on the team.
- *On-site vs off-site impact.* Is the safety risk and/or environmental risk localized; e.g., 1-3 people in a unit affected, or several people in a nearby building affected? Are people off-site at risk? Are there sensitive locations, such as a school, or care facility nearby, are streams and rivers or navigable waterways impacted? Offsite impacts may require a corporate representative.
- *Consequence.* Is the consequence considered a worst case or maximum credible event? Are the consequences reversible or not? Is this a High Consequence Low Probability (HCLP) event that requires special consideration? (Many companies have special decision processes required for HCLP or catastrophic events).
- *Likelihood.* Is the initiating event considered to be a very low frequency one (i.e. outside the range of normal human experience)? Then a more structured, quantitative frequency estimation (like LOPA) may be more appropriate. If an actual incident is the origination of the risk decision, the team, and organization, may need to review its thinking about the likelihood of certain events.) This affects the decision tool selection and may lead to the need to have an expert with that tool on the team.

- *Business disruption.* Is the business loss expected to be minor with quick recovery (one or a few equipment items), moderate unit damage (several equipment items), or plant-wide damage (units/buildings damaged)?
- *Corporate Image.* How concerned is the company about its public image? Would a major event potentially impact the company's reputation and its ability to conduct business or grow its operations?

As the consequence or risk increases, the higher up in the organization level the decision may need to go. The team should have a member that represents the organization, or at least establish a communication plan, with the appropriate organizational representative. Further, in some cases, input from the corporate legal department may be prudent as some complex risk assessments may create potential liabilities upon discovery.

Approval of complex risk decisions must be elevated from the team to senior level approval teams including multidiscipline members before actions are taken. The issue of *groupthink* (see Section 6.8 - *Groupthink Trap*), a phenomenon in which poor decisions can be made because people subordinate their opinions to that of a group to promote consensus and avoid conflict, can be reduced using this approach. Quantification, either semi (e.g. LOPA) or full (e.g. QRA) can moderate groupthink issues. Complex risk decisions, especially those involving severe consequences, should not be made by a single person.

Case study. A company needed to expand capacity for a line of specialty chemicals. The business responsible for the project had manufacturing capacity in two plants and was interested in consolidating all production into one plant. Plant A needed extensive modification to retrofit the process into the existing unit. The plant engineering group would oversee the project. Plant B required a new production line. The company's engineering division would oversee the project.

The assembled team consisted of the unit engineer from each plant, a research engineer from the business' research group, a process engineer from the corporate engineering group, and an economic analyst. No process safety representative was involved initially because the same process hazards would be present for each option. The team reported to the business manager. Capital cost favored Plant A; operating costs favored Plant B. The cash flow analysis showed no clear "winner", and the decision seemed a toss-up. The deciding factor came down to a belief that the operational benefits of a plant designed to run the specialty chemical processes from the start, as opposed to retrofitting equipment into an existing unit, would lead to smoother operation, and potentially better quality. The team also thought that potentially inherently safer design options might emerge as detailed design started.

5.3.2 Opening Meeting

An opening, or kickoff, meeting is needed for intermediate and complex decisions that characteristically involve people with different skills, backgrounds, and outlooks on the problem. In addition to introductions, it is at this meeting that the nature of the problem is described. Input from the person with ultimate responsibility for the decision is very likely. Even if the ultimate decision maker initially defines the problem, the team should still review the problem definition later to avoid the framing trap described in Section 6.6.

In addition to the general outline of the problem the initial input can include restraints such as a deadline for the decision, resources available to the team, capital cost limits, etc.

If a team is using the processes described in this book, a review of the decision process, outlined in Section 4.3, and decision traps, described in Chapter 6, is a good way to start the process after introductions and problem overview. If the decision involves an existing process unit or plant, a tour of it may be in order. (In the case study above, the team toured both plants during the kickoff meeting. As a result, that meeting lasted two days.)

At the kick-off meeting (or perhaps a second meeting if needed), the team can define the problem (Section 4.3.1) and identify the alternatives (Section 4.3.3). Depending on the time involved, however, the team can opt to do this at a second meeting. As the problem and alternatives are defined, the tools needed to screen alternatives will become apparent. At this point it is a good time to reassess the team make-up and determine if other people are needed. For example, if the team thinks a more complex risk analysis tool is needed, e.g. FTA/ETA or QRA, an expert in the use of these tools can be brought on the team or used as a resource. The team will need to decide this. The team should also bring problems with making the decision, such as insufficient time and resources for the work needed (perhaps an external subject matter expert is needed to do a QRA) to the attention of the final decision maker.

At this time the team may also decide what decision method they expect to use, such as an objective/outcome table, simple ranking, weighted scoring, or other (Sections 4.4 and 4.5).

5.3.2 Tools/Methods

An organization should use a consistent set of tools, or hierarchy of tools, going from simple to complex, with guidelines on when to go up in complexity. If an organization has elected not to have certain in-house expertise for use of more sophisticated tools, such as FTA or QRA, outside technical expertise may be needed. There are many consulting firms that specialize in calculating process safety risks. They may use slightly different tools for their analysis. An organization that contracts these third-party firms should make sure the contractors are aware of the organization's risk and decision criteria and that input data and assumptions for different studies are made in a consistent manner.

5.3.3 Time

As the tools and decision become more complex, more time will likely be needed to complete the analysis. The organization must plan to allow the people working on the decision enough time to do the analysis and make the selection. Even if a study is done by an external consultant, internal resources must be available to provide input, answer questions and review the results. Therefore, any time constraint should be considered in terms of people's priorities and the type of tools used.

5.4 DEFINE DECISION CRITERIA

5.4.1 Process Safety Risk Criteria

Section 2.4.2 described qualitative and quantitative risk criteria. The individual or team charged with the risk decision needs to decide which risk measure is appropriate to the decision. The criterion for the other objectives (Section 4.4) also needs to be determined. In some cases, these may be already defined by an organization, in others the decision team needs to define those objectives.

With *consequence-based* criteria, an organization establishes the severity [or consequence] criteria based on Loss of Primary Containment (LOPC). A consequence-based criterion (Section 2.3.1) can be qualitative criteria; High, Medium and Low. For example:

- High – a large flammable or toxic release that can cover/impact a site or facility (and/or off-site), for example a rupture of an equipment item, or large pipe leak or leak that goes undetected for a long period
- Medium – a moderate sized flammable or toxic release that could affect a unit, for example a leak due to failure of a flange, relief valve opening, etc.
- Low – a small release/leak that affects only the immediate vicinity

The consequence-based criteria can also be quantitative, with release rates or sizes specified. Criteria based on an estimated amount of material released based on its chemical and physical properties, such as flammability and toxicity can be established. Table 5.1 from *Layer of*

Table 5.1. Example Consequence-Based Consequence Categorization (Releases)

Release Characteristic	Size of Release, pounds beyond a dike and category					
	1 to 10	10 to 100	100 to 1,000	1,000 to 10,000	10,000 to 100,000	>100,000
Extremely toxic, above BP*	3	4	5	5	5	5
Extremely toxic, below BP or highly toxic, above BP	2	3	4	5	5	5
Highly toxic, below BP or Flammable, above BP	2	2	3	4	5	5
Flammable below BP	1	2	2	3	4	5
Combustible Liquid	1	1	1	2	2	3

*BP – Boiling Point

Protection Analysis (CCPS 2001) provides an example of safety consequence-based categories for a hazard assessment or a risk matrix, based on the size and hazard properties of a release. Table 5.2 provides an example of business consequence categories. In these tables, the worse the consequence, the higher the number assigned to it. The numbers themselves do not have any significance, some companies use roman numerals (I – V) or letters (A – E).

With *impact-based* criteria the organization establishes criteria in terms of impact, such as the numbers of fatalities, level of environmental impact, or financial loss due to equipment damage or loss of production that can be tolerated for each scenario or for a unit or facility. A An impact-based criterion can be qualitative, relying on judgment, e.g. High, Medium, or Low, but is more likely to be at least semi-quantitative.

When done qualitatively, as in the risk matrix in Figure 2.2, judgement is needed to assess probabilities of the impacts for either consequence or impact-based criterion.

If done quantitatively, impact-based criteria require more calculations than a consequence-based criterion since factors such as probability of exposure, probability of ignition, and probability of fatality given an exposure might need to be determined. However, they do allow

Table 5.2. Example Business Impact Consequence Categorization

Consequence Characteristic	Magnitude of Loss and Category					
	Spared or nonessential equipment	Plant outage <1 month	Plant outage 1–3 months	Plant outage >3 months	Vessel rupture 3,000 to 10,000 gal 100 to 300 psi	Vessel rupture >10,000 gal >300 psi
Mechanical damage to large main product plant	2	3	4	4	4	5
Mechanical damage to small by-product plant	2	2	3	4	4	5
Overall cost of event	0 -10	10 - 100	100 – 1,000	1,000 – 10,000	> 10,000	
Category	1	2	3	4	5	

to take local conditions into account. Some actual quantitative risk criteria are presented in Chapter 10.

5.4.2 Other Criteria

Other decision criteria were discussed in Section 4.4 – *Objectives and Outcomes*, and in the Section 5.3 – *Assembling the Team*. These criteria may have to be established on a case by case basis, unless defined on a corporate wide basis.

5.5 MAKING THE DECISION

5.5.1 Characteristics of Decision Aids

Decision aids were discussed in Chapter 4. A few decision aids were introduced in Section 4.1. Decision aids, in order of increasing complexity, are summarized below.

Engineering Judgement. Competent hazard assessment teams, engineers or operating personnel are capable of making good decisions based on experience.

Standards/RAGAGEP. Third party standards, consensus RAGAGEPs, and internal RAGAGEPs are documented good practices based on the experience of subject matter experts in specific technical areas. Hence, they represent good engineering judgement written down over time so others with less experience don't have to learn the hard way.

Objective/Outcomes Table. This table is described in Section 4.4. The table contains the objectives of the decision, and the outcomes of each alternative with respect to the objectives. In many cases, a review of this table will enable a good decision to be made, or at least eliminate some alternatives.

Simple Ranking or Weighted Ranking of Objectives/Outcomes. When a simple listing of the objectives and outcomes does not lead to a final decision, the remaining alternatives then have to be evaluated by some kind of ranking system. Developing the rankings involved judging how important the outcomes are relative to each other, i.e., making trade-offs.

5.5.2 Appling the Decision Tools, Aids, and Criteria

The first step in applying a decision tool is to define the nature and complexity of the problem and alternative solutions. Next the decision criteria should be determined. Three examples, representing low, moderate and high complexity decisions are presented below as examples.

Quick and/or simple decisions. The scenario is an overflow of flammable material from a tank due to failure of a level control loop. The risk is that flammable vapors from the overflow materials find an ignition source and result in a flash fire or explosion. This could cause an injury or fatality, especially if an operator is present in the impact area. (In real life, this seemingly simple scenario has occurred at gasoline storage terminals in Buncefield, U.K., and at a CAPECO facility in Puerto Rico, causing large explosions and severe damage to the facilities.) The alternatives could be a hardwired high-level switch, a high-level shutoff SIF, or operator response to a high-level alarm independent of the control loop. Tools that would be applicable to this situation would be application of a pre-specified design code (RAGAGEP), or a risk matrix. Each of these approaches would likely lead to the choice of multiple protection layers, including a high-integrity, high-level shutoff (e.g. a SIF), each with increasingly more rigorous analysis.

Intermediate risk decision. The concern is an exothermic runaway reaction in a semi-batch reactor where one reactant is co-fed with a catalyst to a vessel with the other reactant at a specified temperature range. Temperature in the reactor is maintained by control of the feed rates and application of cooling to the reactor. The runaway reaction can overpressure the vessel. This could destroy the unit and cause multiple fatalities. A PHA team may find several possible scenarios:

- Loss of or inadequate cooling
- Failure of the feed control loop (includes a temperature sensor, mass meters and control valves) leading to high flow of the reactant and catalyst
- Loss of agitation (leads to accumulation of unreacted material which then reacts after enough build up)
- Failure of the feed control loop leading to low temperature (leads to accumulation of unreacted materials which then react all at once)
- Low or no catalyst flow (leads to accumulation of unreacted materials which then react all at once)
- Poor quality catalyst (e.g. contamination or past recommended shelf life)

There is an emergency relief system (ERS) that can safely relieve the runaway reaction of the high temperature scenarios (the first two failure modes above), but which leads to a release of a toxic by-product. The relief system cannot safely relieve a runaway reaction scenario of build-up of more than 25% of the feed without the expected reaction starting (the last four failure modes). (This could be a characteristic of a nitration reaction, for example.)

There are at least two risk decisions that should be made here:

Thermal runaway scenario: What layers of protection are needed to decrease the runaway reaction likelihood enough in combination with the ERS to meet the risk criteria for vessel rupture? What, if any, treatment is needed for the toxic effluent from the ERS?

After process calculations such as heat of reaction and rate of pressure rise, have been done, LOPA or FTA can be used to define and evaluate alternatives for reducing the likelihood that the potential scenarios can lead to vessel rupture to meet the risk criteria. For the ERS release, the organization can opt to:

- Reduce the likelihood of the release below the risk criteria by adding enough layers of protection to make the likelihood of the release itself below the risk criteria using LOPA or FTA
- Analyze the release consequences and treatment options to determine if the risk of fatality from the release can be reduced by lowering the severity of the consequence to meet risk criteria using a combination of FTA/ETA

Reactant accumulation runaway scenario: What layers of protection are needed to eliminate or reduce the likelihood of the runaway reaction without the need for ERS to meet the risk criteria? Options can be:

- Perform the reaction only in a reactor with a design pressure that can contain the reaction (an example of Inherently Safer Design (ISD))
- Decrease the likelihood of the scenario to a tolerable level (section UG-140 of the ASME Boiler Pressure Vessel Code allows an organization to choose overpressure protection by system design instead of an ERS if the likelihood of the scenario can be reduced to an acceptably low level, as established by the organization)
- Install a larger relief nozzle or second nozzle to reduce the risk of overpressure
- Design a reaction kill system that injects a material that stops the reaction

If the organization opts the last three bullets, LOPA or FTA can be used to find the combination of protection layers to meet the risk criteria. Since LOPA is typically an order of magnitude method, an organization may choose FTA to better identify and quantify the protection layers needed.

Complex risk decision. An example of a complex problem may be one involving a refinery and petrochemical plant. Such a plant will contain many processes, including catalytic and thermal cracking, reforming, isomerization, and polymerization processes, handle flammable materials and product storage (e.g. natural gas, gasoline, propane), and may have toxic materials such as hydrofluoric acid, hydrogen sulfide (a byproduct of many refinery processes), ammonia and chlorine. Many refineries and petrochemical plants are in areas close to residential or mixed-use areas, creating the potential for off-site risk. A large LNG facility would also be an example of this. There are a multitude of risk decisions to make. Some risk decisions may involve interactions between units or recycle streams within a unit (e.g. fertilizer production). Prioritizing risk decisions may be necessary.

One example of a complex decision is whether to use sulfuric acid (H_2SO_4) or hydrofluoric acid (HF) in the alkylation section of a refinery. HF based alkylation uses less acid and that is mostly replenished and recycled. Sulfuric acid alkylation units use larger amounts of fresh acid and a means to recover the spent acid is needed. HF offers some other processing advantages, such as being usable with a wider range of feedstocks.

HF is a highly hazardous material. While sulfuric acid will cause acid burns, emergency response is typical for acids, it can be washed off with water. HF will penetrate skin and muscle and go to the bone if not treated. Also, exposure to an area of skin the size of your hand can be fatal. It cannot be washed off with water and specialized treatment is needed, e.g., with calcium gluconate, to stop the acid from continuing to cause burns. HF requires very special handling protocols. Use of large amounts of sulfuric acid, however, may require transportation risks if the fresh acid cannot be made and the recovered acid cannot be treated in-site. This can expose the public to the hazard of road transportation.

If the reaction example in the intermediate risk decision section is changed from a single purpose, stand-alone plant to one inside a larger facility, the decision can become more complex. An overpressurization of the reactor might now affect adjacent units, causing domino effects. Even a release from the ERS might now affect personnel operating an adjacent unit, forcing an unplanned shutdown, for example. Or, move the plant from a remote setting to one with nearby neighbors, the decision becomes more complex.

5.5.3 Recognizing and Dealing with Uncertainties

Section 4.6 described sources of uncertainty in risk decision tools. When the risk tools are being applied to make a decision, the person or team needs to recognize and deal with these uncertainties.

Incomplete risk definition. If the risk decision comes from a PHA, then a review of the PHA to check its scope, see who the attendees were and if their knowledge was appropriate with respect to the process technology, and an assessment of the overall quality of the PHA is in order. *Revalidating Process Hazard Analyses* (CCPS, 2001a) has checklists for evaluating the quality of a PHA.

A term to watch out for in a PHA is "not credible". In some cases when a team thinks a scenario is not credible, it may really be that its likelihood is so low it is difficult to imagine. This is an example of a decision trap, forecasting and overconfidence. Decision traps are discussed in Chapter 6.

The boundary between what is a very low frequency with what is not possible can be a very fine line (unless the team used not credible to mean physically impossible). It is very difficult, if not impossible, for a PHA team to predict the likelihood of low-frequency/high-consequence events. This is what tools such as LOPA, FTA, ETA and QRA exist to do. To repeat, when "not credible" appears in a PHA, an alarm should go off in the heads of the risk study team.

Lack of process knowledge when doing a PHA is a symptom of an inadequate team. Very few process incidents involve a totally new and previously unidentified cause such as a new chemical mechanism, or unknown metallurgical issue, for example. Comparing the PHA team's background and qualifications to the hazards of the process is one way to assess if the team has the right process knowledge.

Adequate time and resources like trained manpower and PHA team leader play essential role in PHA. If the risk analysis team has concerns about the PHA, the team members, or PHA quality, consultation with other process and/or technology experts with the appropriate knowledge can be done to look for gaps.

If the decision maker or team believes there is a shortcoming in the scope, they can check for other PHAs that were done on areas outside the subject PHAs scope. Examples would be utilities used by the process, or processes upstream and downstream of the source PHA that might impact it.

Residual risk. With simple decision problems in which only a few scenarios are being analyzed, residual risk is not a major issue. With intermediate or complex problems in which many scenarios are being analyzed, there is a chance that the implementation of recommendations may have to be prioritized, with some being implemented quickly and others taking longer or having to wait for an opportune time, such as a process shutdown. In such cases, the risk from those scenarios needs to be called out and interim, sometimes administrative, solutions may need to be implemented.

One of the advantages of quantitative tools such as FTA, ETA and QRA is that the contribution of each scenario to the overall risk can be determined. Residual risks are identified, so the organization can target them for future risk reduction and can avoid making changes that might increase the risk from these scenarios.

Frequency data. A study team and decision maker need to be aware of the nature of the frequency data used when making risk decisions. If the data cannot be based on statistically valid plant data, the data used in lieu of that may have a large margin of error (perhaps an order of magnitude).

Caution: If a risk tolerance criterion is $1x10^{-4}$/year and a result of $8x10^{-5}$/year is calculated, the organization should not consider this a precise answer and assume its work is done. On the flip side, do not be surprised if a calculated risk result is $1.2x10^{-4}$/year, and someone asks if that is an acceptable result because it rounds down to $1x10^{-4}$/year. In either case, when a result is near the risk criterion line, and if further risk reduction can be done at a reasonable cost, it should be pursued. This is an example of the application of ALARP.

Consequence analysis. When modeling releases of materials, some companies can decrease the concentration of concern to some fraction of the normal value. For example, for a flammable release with fire and explosion consequences, a fraction of the Lower Flammable Limit, such as 50% or 25% of the LFL, can be used to account for the potential margin of error.

QRA. QRA studies, which employ both frequency and consequence analysis, tend to have moderate to high uncertainties and care is required to address these uncertainties in decision making. One way to account for the uncertainty is to have a band of tolerable risk. This is similar to using ALARP. The upper end of the risk band could be tolerable but further risk reduction should be pursued and the lower end would be a region below which the risk is acceptable.

Applications of techniques that use frequency data and consequence analysis are best used when evaluating multiple options. Even though there is uncertainty in the final risk measure of an option, the ranking of different options analyzed by the same techniques should still be correct. This is the preferred use of QRA in decision making.

5.5.4 Recognizing the Need to Escalate the Decision

As the outcomes of a risk decision increase, the more people and assets the decision impacts. A decision that can affect not only the immediate locale, but an entire plant, or even off-site populations should be made at the appropriate plant, business and organization levels. Companies should consider establishing guidance on when legal input should be considered as part of the risk decision making process. The organization should establish guidance on when to escalate a decision. Possible levels and positions involved could be:

Immediate area. At the production floor, operators, supervision, and technical support (production engineers) need to be trained on when to call for extra technical support before continuing. This guidance can be in the troubleshooting guide as well. The technical support personnel need enough training and experience to either make a decision, or if the risk is high and there is uncertainty about the right thing to do, to stop and call for more expertise if needed.

Entire unit. At this level, an operations manager, business manager or representative, and perhaps the facility manager should be involved. Plant HSE managers should also be included in the approval level.

Entire plant. At this level, all of the above plus a corporate manager should be involved. If the plant has units that support several businesses, this person must represent the needs of the other businesses affected.

Off-Site. All of the above plus a corporate HSE and perhaps a higher level corporate executive can be involved.

Companies should also establish guidance on when legal input should be considered as part of the risk decision making process.

5.6 FINALIZING DECISION AND THE APPROVAL PROCESS

Once the decision is made, the person or team making the decision should inform the management or sponsor of the decision and the rationale behind it. The higher the capital cost or more significant the changes in the operating process, equipment, and procedures are, the higher the level in the organization the approval level will be. Some organizations may require that risk decisions that affect other operations in the plant or have off-site impacts also go up higher in an organization. In Section 5.3.1 – *Team Members*, the need for including a representative of the highest decision-making level was mentioned. That representative can keep the decision-maker in the loop to avoid surprises.

For example, a decision affecting risk in a unit may be approved by a plant or site manager. A decision affecting larger segments of a plant or causing business interruption but with no off-site risk impact may require approval from a site and business manager. A decision affecting the plant and off-site populations may need plant, business and corporate management approval. Such a decision may require approval from other corporate units such as a health, safety and environmental group, a quality control group (in the case of FDA or CGMP), and/or a legal group. A team member who keeps the decision makers in the loop, as mentioned in section 5.3.1, can make the approval process easier.

The overall relationship between risk decision categories, risk analysis tools, decision aids, and approval level is summarized in. Table 5.3.

5.7 COMMUNICATING, DOCUMENTING, AND IMPLEMENTING THE DECISION

Documentation of the decision should:

- Clearly state what the problem's hazard and existing risk was and what the decision is for the path forward
- Include any recommendations made as a result of the decision, including alternative solutions and rejected recommendations and the basis for rejecting them
- Define the scope and limits of any studies used in the decision
- List assumptions made in the study
- Contain enough information to allow scrutiny of the decision

Table 5.3 Decision Level, Risk Tools, Decision Aids Summary

Decision Level	Quick and/or Simple	Intermediate	Complex
Risk Analysis Technique	Engineering or PHA Team judgement Qualitative Risk Matrices	Semi-Quantitative Risk Matrices LOPA Consequence Modeling FTA	LOPA Consequence Modeling FTA ETA QRA
Decision Aids	Single scenario risk criterion Engineering Judgement Voting/Consensus Checklist Standards and/or RAGEGEP	Single scenario or unit level risk criteria Standards and/or RAGAGEP Objectives / Outcome Table Outcome Rankings	Unit / facility and company level risk criteria Objectives / Outcome Table Outcome Rankings Weighted Scoring
Approval Level	Local (Individual or unit)	Plant or business unit	Business unit or corporate level

Quick and/or simple decisions. A quick decision needs to be communicated to the appropriate levels quickly. For example, response to a process deviation may imply something went wrong, and unit supervisors need to know as soon as possible. A method to communicate a simple decision problem, could be to describe the scenario or scenarios, the unmitigated risk of each and the reduced risk or the Risk Reduction Factor after the recommendation is implemented. If a risk matrix was used, then the risk category with and without the recommendation can be provided.

Intermediate decisions. A table listing of the decisions and their impact on the risk, as described for simple risk decisions can be used for medium complexity decisions as well. Another method to communicate and document the findings of a moderately complex risk study is a Bow Tie (also called barrier) diagram. The diagram communicates the results graphically (see Figure 2.6). This can enhance the understanding of the results.

Complex decisions. A table of the risk reduction of each finding or recommendation on the system's overall risk, as well as the cumulative effect of the recommendations can be used for complex decisions. If a QRA is performed, the risk contours and F-N curves (see Figures 10.5 and 10.6) of the system before and after implementing the recommendations should be presented.

All groups affected by the decision should be informed of and trained on its consequences and implications, as necessary. Such communication/training methods may include verbal communication from supervisors, formal training sessions, change notices documented in procedures, entries in logbooks, written summaries of changes, e-mail notifications, and other approaches. This includes any assumptions made about the operation during any risk analysis work going into the decision. In some cases, detailed training may be necessary. Such training will enable the appropriate personnel to understand how changes or new information that comes up in the future affect the decision and its results. Documentation is also needed to prevent loss of process knowledge.

5.7 SUMMARY

One of the twenty elements of the CCPS RBPS management elements is Process Knowledge Management (CCPS 2007). The goal of knowledge management is to understand the risks of a process. A risk decision, and the inputs into it, becomes part of the process safety knowledge and should be recorded. Future PHA revalidations and MOC reviews should incorporate any knowledge gained as part of any study done for the decision. In the US, for processes covered by the OSHA PSM and EPA RMP regulations, the equivalent of Knowledge Management is the Process Safety Information (PSI) requirement. These regulations require a written compilation of PSI, specifically written information on the hazards of the chemicals used or produced, and on the technology and equipment used in covered processes. Any risk decision that affects the PSI should be documented, even if the process is not covered by OSHA PSM. Local authorities and other countries are likely to have similar requirements. The book, *Guidelines for Process Safety Documentation* (CCPS, 1995) provides examples of how to document PHAs, quantitative studies, etc.

Implementing the risk decision will, by its nature, be a change to the process operation in some way. An MOC review, and very likely a Pre-Startup Safety Review (PSSR) will be needed when implementing the decision.

6

POTENTIAL DECISION TRAPS

6.1 INTRODUCTION

"The fault, dear Brutus, is not in our stars, But in ourselves, that we are underlings."[2]

Previous chapters dealt with the decision process. The uncertainties that could affect the decision were related to factual issues such as scenario definition, frequency data, and methodological issues were covered in Section 5.5.2. This chapter is about the human side of the decision, the decision maker and team. Complex decisions involve many inputs, sometimes coming from different people or stakeholders, each with their own biases and assumptions. There are several traps that a decision maker can fall into because of these biases. These are described in *The Hidden Traps in Decision Making* (Hammond, Keeny and Raiffa, 1998) and in other sources. The decision traps that will be described in this chapter include:

- The Anchoring trap
- The Status-Quo trap
- The Sunk-Cost trap
- The Confirming-Evidence trap
- The Framing trap
- The Estimating and Forecasting traps
    - The Overconfidence trap
    - The Prudence trap
    - The Recallability trap
- The Groupthink trap

The best defense against these traps is to be aware of them and how they affect your thinking. They are discussed in this chapter with respect to decisions about risk reduction.

6.2. ANCHORING TRAP

Hammond, Keeny and Raiffa (1998) state, "Initial impressions, estimates, or data anchor subsequent thoughts and judgements." We tend to give too much weight to past history or the first information we receive.

---

[2] Shakespeare, W., Julius Caesar, Act 1, Scene 2.

6.2.1 Anchoring Trap Example, Titanic

A good example of anchoring is the tragedy of the Titanic, April 14, 1912. The Titanic sank after striking an iceberg. Decisions made before, during and after the impact with the iceberg were based on the belief that the Titanic was "unsinkable". Everybody on board, including the ship's captain, a managing director of the White Star line (owners of the ship), and the passengers, were told the ship was unsinkable many, many times.

Everyone was anchored in the belief of unsinkability; all risk decisions were affected by this belief. To save deck space, fewer lifeboats were provided on board than what were needed to hold all the passengers and crew. After all, the ship was unsinkable. Prior to hitting the iceberg, the Titanic was traveling at full speed, at night, through an area known to have icebergs. After all, what could happen? After the iceberg was struck, the ship stopped. While stopped the initial flooding could have been controlled long enough for other ships to come and render aid. Good seamanship would have been to stay stopped until a full damage assessment was finished, but the owner pressed for the voyage to continue. The captain gave in to the owner's pressure to restart the voyage because the belief that the Titanic was unsinkable was too strong. Finally, many of the lifeboats that were lowered were not filled to capacity with people because the passengers also thought the ship unsinkable (Would you get off the biggest ship ever built, which you *know* is unsinkable, and to go into the North Atlantic in a tiny boat?) (CCPS 2008a).

6.2.2 Countering the Anchoring Trap

When making risk-based decisions, it is possible, even likely, that the event that is being assessed hasn't occurred. As such, there is a bias to assume that because the events haven't happened, they won't happen. Decision makers can illustrate the anchoring bias and may not accept the risk decision that the process is recommending. This can be particularly challenging for low frequency, high severity events where people will struggle to see the possibility of the scenario. This situation creates a challenge when trying to get people to address unacceptable risks which have been identified through a risk-based decision process.

To guard against the anchoring-trap:

- Ask, are you weighing all data and calculations equally? If not, why? Ask subgroups to look at the data separately
- Ask different team members to look at the decision from different perspectives
- When asking for input from outside the team, avoid stating your own opinion so as not to anchor them

6.3 STATUS-QUO TRAP

The Status-Quo trap is a "strong bias toward alternatives that perpetuate the status quo" (Hammond, Keeny and Raiffa, 1998). In a process plant change may be expensive, increasing this bias. With limited funds the idea of spending money on something that is perceived to be an acceptable risk or is not value added can be especially hard.

6.3.1 Status Quo Examples

BP Refinery Explosion Texas City, 2005. The 2005 Texas City BP refinery explosion provides one example of the status-quo trap. The flammable release was through a blowdown stack that was open to the atmosphere (CSB 2007). The stack was installed in the 1950s. By the 1970s, release to the open air was no longer considered a good practice. In fact, Amoco (the original owner of the refinery) had a process safety standard that required existing atmospheric blowdown stacks to be phased out and prohibited new ones from being built. In a 15-year period before the incident, there were 3 proposals to replace the blowdown stack, and there was an OSHA citation in 1992 which recommended changing the vent to go to a flare as a way to comply with an OSHA finding. None of these proposals were acted upon. The plant was locked into the status-quo trap and did not perceive a driving reason to change.

West Fertilizer Fire and Explosion, West, TX, 2013. West Fertilizer stored ammonium nitrate prills (a prill is a pellet or solid globule formed by the congealing of a liquid). The fire and explosion itself are not an example of the status-quo trap (a lack of hazard awareness was a key factor). The severity of the consequences are an example of how the status-quo trap can work. When the facility was built (1962), it was located in an area with open fields. Over the intervening years, the town of West, TX, grew. (The CSB report has a series of photos that shows how the area around the facility grew (CSB 2013, p. 229).) By the time of the incident, an apartment complex was located about 450 feet (137 m) from the site, a nursing home about 500 feet (152 m), an intermediate school about 550 feet (167 m), and a high school about 1,300 feet (396 m). Two people at the apartment complex were killed and all of these buildings were so badly damaged they had to be demolished (CSB 2013).

The status-quo, in terms of the site's location relative to population, had changed. This has actually happened to many process facilities. Not adjusting to the change in the off-site population because the facility is thought to be "safe" based on previous studies is an example of the status-quo trap.

6.3.2 Countering the Status-Quo Trap

What was once viewed to be safe may no longer be considered safe. Times change, and along with time, societal expectations also change. As such, outdated designs need to be looked at very closely and the status quo needs to be challenged.

One approach for addressing the Status Quo trap is to remind the team and decision makers about the changes that have occurred in automotive safety over a similar time period.

> Modern vehicles have many more safety features than a vehicle manufactured in the 1950's (when the Texas City Refinery installed the blowdown tower). The prevailing attitude was, if you were hurt in an accident that was your own tough luck. When seatbelts were introduced, many people resisted wearing them. This author (who is now in his late 60's) can personally remember people saying, "I'd rather be able to get out of the car". In fact, data showed it was safer not to be thrown out of the car. Today, vehicles have 3-point seatbelts that alarm if not buckled, and interior designs and air bags to protect occupants in case of a crash. Also, cars are designed to absorb the energy of impact by collapsing (which is why a seemingly small accident can cause thousands of dollars in damage, an example of the expression "no good deed goes unpunished"), thus further protecting the occupants. Auto manufacturers now use safety ratings and safety systems such as lane departure alerts and automatic braking as a selling point.

To counter the status-quo trap:

- Remember that the perception of tolerable risks can change with new knowledge or changing social expectations
- Ask if you would choose the status-quo option if it did not already exist
- Remember the objectives of the decision
- Identify and rank the risks for as many alternatives as possible and remember what they are

6.4 SUNK-COST AND ESCALATION OF COMMITMENT TRAP

Sunk costs are resources already expended on a project such as research, engineering design, capital investment, or other items. People tend to try to justify their previous decision or expenditure and can be biased in their choice of options towards the one that resources were already expended upon. When a decision is made, it should be based on what is the best alternative from that point onward.

Consider this hypothetical example: A plant has just invested $100,000 to upgrade the process control system of a unit or a unit operation. A risk study now finds that the unit needs to install a Safety Instrumented System (SIS) to enable the unit to meet risk criteria. This will involve a capital investment and continuing operating costs for the testing and maintenance that will be required for the life of the plant. An alternative is proposed that is inherently safer. It will require more capital initially but will have lower operating costs and is more sustainable. Human nature is to favor choosing the SIS because of the previous investment. What the organization should do is perform a Net Present Value (NPV) calculation for the two alternatives and compare their Risk Reduction Factors (RRF) along with the NPV. The initial $100,000 upgrade should have nothing to do with the decision.

6.4.1 Countering the Sunk-Cost Trap

To counter the sunk-cost trap, ask yourself, if new leaders were put in place, what would those new leaders do? This mindset can be helpful as it can allow decision makers to let go of past decisions and to focus only on the new decision to be made. To counter the sunk-cost trap:

- Consciously ask yourself if you are being driven by the sunk-cost syndrome. Be on the lookout for that mentality.
- Remind yourself that the sunk-cost trap exists, and it is a fallacy to use it.
- Remind people that a new decision does not mean the previous expenditure was a mistake. It was probably correct at the time, and that new information does not change the past.

6.5 CONFIRMING-EVIDENCE TRAP

Confirming evidence is the bias to seek out data or information that will support an existing belief. In a risk decision this can take the form of a decision maker (e.g. a plant manager or design engineer) trying to decide how likely a risk scenario might be. It has never happened in their experience, so they think it very unlikely. To help them decide, they ask another plant manager or design engineer if they ever heard of it happening. That person is likely to confirm their belief, especially if they share the same experiences, training, etc., and are a colleague of the decision maker.

Another form of confirming evidence is deliberately looking for information to confirm existing belief rather than testing the belief. This can take the form of deliberately asking someone who you know will agree with you for an opinion.

6.5.1 Countering the Confirming Evidence Trap

A way to guard against the confirming bias is to seek out input from a wider group of people. In process safety, this also means seeking out incident and near miss data from outside one's own experience.

A corporate database of incidents can help with this, or a review of case history literature can be conducted.

> The CCPS publishes a free monthly one-page Process Safety Beacon (PSB). Each issue presents a real-life accident, and describes the lessons learned and practical means to prevent a similar accident in your plant. Subscribing to and widely distributing the PSB within a plant can help overcome the confirming evidence trap.
>
> CCPS also runs a database of incidents collected from a consortium of companies who are members of the database (the Process Safety Incident Database) which has over 800 incidents described in it. This allows companies to find evidence beyond their own internal experience.

In a team setting, one or more team members can be assigned a role of devil's advocate to challenge the evidence.

When determining the objectives (Section 4.4), make them as quantitative as possible to reduce subjectivity in the outcomes.

6.6 FRAMING TRAP

"People did not choose between things. They choose between descriptions of options" (Lewis, 2017). How a problem is stated can affect the decision. Poor framing of a question can cause anchoring or a status quo bias. For example:

1. A process modification to reduce risk in a new plant will add $1,000,000 to the capital cost. Do you make the investment?
2. A process modification to reduce risk in a new plant will add $1,000,000 to a $50,000,000 capital project. Do you make the investment?

If you are like most people, the first question seems like a harsher choice, because you are comparing it to $0. In the second question, the investment is put into context with the total capital cost and seems more reasonable.

6.6.1 Framing Example

People tend to be risk averse. Take this hypothetical example of framing a facility siting decision. A study team is trying to decide between two options, Options A and B. 600 people are affected by the potential release. Option A could potentially result in 400 deaths. Option B has a 33% chance no one will die, but a 67% chance of 600 fatalities. Table 6.1 shows two ways to frame the decision.

Most people will choose Option A when presented with the positive framing. Conversely, most people will choose Option B when presented with the negative framing.

6.6.2 Countering the Framing Trap

Deliberately restating a decision problem in different ways can help a decision maker determine if framing of the question is influencing their outlook. If a study team is presented with a decision problem, they do not have to accept the initial statement. A person or team can avoid the framing trap by restating the problem in different ways, perhaps by presenting options in both a negative and positive frame so a decision maker is made aware of the framing issue and focus on the data.

6.7 ESTIMATING AND FORECASTING TRAP

Many high consequence events are also low frequency events. Understanding low frequency (e.g., an event has a likelihood of 1/1,000,000 per year) is difficult for most people.

6.7.1 Overconfidence

People can place too much confidence in previous history and be overconfident in the accuracy of their judgement. An example of this is the Columbia space shuttle loss in 2003.

Overconfidence Example, Columbia Space Shuttle. On February 1, 2003, the Columbia Space Shuttle disintegrated on reentry due to the damage to its thermal tiles caused by impact with insulating foam shedding from the fuel tank during take-off, Figure 6.1. The impact was detected by cameras during the ascent, and a Debris Assessment Team was formed to analyze the potential damage.

Table 6.1. Framing example

Framing	Option A	Option B
Positive	Saves 200 lives	33% chance of saving 600 lives
Negative	400 people die	66% chance of saving no lives

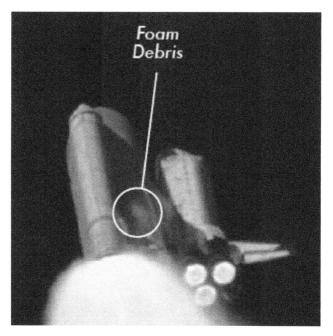

Figure 6.1. Foam debris striking shuttle Columba. Courtesy NASA.

This team requested that images be obtained from orbital satellites several times (the official report documented eight times). These requests were refused, and the team instructed to use mathematical modeling to analyze the impact. Modeling was inconclusive. Throughout the mission, engineers from contractors and NASA were more concerned about the potential damage than the NASA mission managers. NASA decision makers eventually concluded there was little risk and decided to continue the mission as planned.

Why? Overconfidence. Most missions had insulating foam shedding during ascent. There was a history of missions with damaged thermal tiles that caused no problems. Indeed, there were presentations and data that confirmed their beliefs, see the conclusion in the slide in Figure 6.2 (CAIB 2003). In the process safety field, this is also called "normalization of deviation" (The reader is referred to *Recognizing and Responding to Normalization of Deviance* (CCPS 2018a) for more about this important phenomenon.)

In decision terms, due to past performance, the mission management was overconfident in the belief that foam loss and tile damage was not a problem and therefore did not respond to the concerns of the engineers

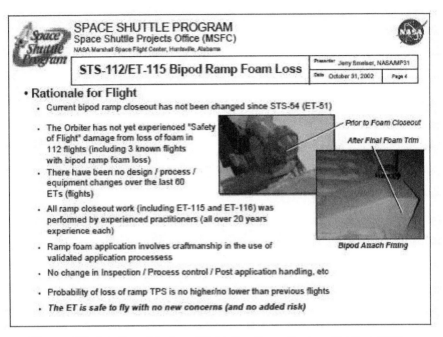

Figure 6.2. Briefing slide from Flight Readiness Review, October 2002, (courtesy NASA).

Overconfidence Example, Fukushima Power Plant Explosion. The Fukushima nuclear power plant was designed to withstand a magnitude 8 earthquake. At the time, the plant was commissioned in 1971, Japanese scientists did not think a category 9 earthquake was credible. Before the plant was built, however, there were two magnitude 9 earthquakes along what is known as the Pacific "ring of fire": in Chile in 1960 (magnitude 9.5) and Alaska in 1964. (There were two more after the plant was built: in Sumatra in 2004 and Chile again in 2004 (magnitude 8.8)).

Some improvements were made to the Fukushima plant's original design based on a study using a methodology developed in Japan in 2002, but these improvements were still not enough for the actual event (International Atomic Energy Agency (IAEA), 2015).

On March 11, 2011, a magnitude 9 earthquake occurred off the coast of Japan. The largest wave from the tsunami caused by the earthquake was 15 m high; the plant's barrier was 5 m high. The resulting flooding damaged the on-site back-up power generators and cut off the main power supply to the plant. This resulted in the loss of cooling for the three operating reactor units, three shutdown units (nuclear reactors generate heat even when shutdown) as well as at the spent fuel pools. The reactor cores in the three operating reactors overheated, the nuclear fuel melted, and the three containment vessels were breached. Hydrogen was released from the reactor pressure vessels, leading to explosions inside the reactor buildings in the units, damaging structures, equipment, and injuring personnel.

Radionuclides were released from the plant to the atmosphere and were deposited on land and on the ocean. People within a radius of 20 km of the site and in other designated areas were evacuated. By December 2011, 23,000 people had been involved in emergency operations. No one was harmed immediately; however, 180 received doses above the occupational threshold limits for radiation. Due to the long-term nature of radiation exposure, it may be decades before the full effects of any exposures are known.

The IAEA report's recommendations (IAEA 2015, p. 4) included some that specifically address the issue of overconfidence:

- "The assessment of natural hazards needs to be sufficiently conservative. The consideration of mainly historical data in the establishment of the design basis of nuclear power plants is not sufficient to characterize the risks of extreme natural hazards. Even when comprehensive data are available, due to the relatively short observation periods, large uncertainties remain in the prediction of natural hazards.

- The safety of nuclear power plants needs to be re-evaluated on a periodic basis to consider advances in knowledge, and necessary corrective actions or compensatory measures need to be implemented promptly.

- Operating experience programmes need to include experience from both national and international sources. Safety improvements identified through operating experience programmes need to be implemented promptly. The use of operating experience needs to be evaluated periodically and independently."

These recommendations apply to chemical plants, refinery and offshore operations as well, and they apply to chemical and operational hazards as well as natural hazards.

6.7.2 Prudence

Prudence is the opposite of overconfidence. Over conservatism can occur in a risk study by using the most conservative, or worst-case consequence modeling results, without taking the frequency into account (or by consistently rounding up frequency estimates). It is up to the people doing the risk study to guard against this. It is possible that overly conservative estimates can result in too many layers of protection and a plant that is harder to operate, and so expensive it is non-competitive.

> Case Study: A facility conducted a facility siting study (FSS) to evaluate consequences of fires and explosions, and the resulting vulnerability of persons in a control room. Based on consequence

results only (intolerable blast damage from a low probability event), the facility decided to spend $3 million dollars on a control room structural upgrade.

If the facility had spent $25,000 on a quantitative risk analysis (QRA) that included consideration of probability of events, they may have concluded the risk was within company tolerance, societal norms, and the expenditure of $3 million was unwarranted.

The $3 million may have been better spent on other risk reduction measures such as training, improvement of operating procedures, mechanical integrity inspections, etc., if looking at a risk tolerability view. Alternatively, the $3 million may be a worthwhile expenditure from a corporate viewpoint.

Basing a decision on an FSS only may be overly conservative. Obtaining additional valuable information through a QRA to support decisions may reduce over-conservatism. All viewpoints need to be considered.

6.7.3 Recallability

People rely on estimates based on the events they remember or have heard about. This can lead to two extremes. One is thinking a scenario is more likely than it really is because of the recent past, but the usual extreme is thinking a scenario is less likely because "it has never happened here".

As Trevor Kletz stated, "Organizations have no memory" (IChemE 2018). New managers and engineers may not know of famous (or infamous?) events such as the Bhopal toxic release in 1984 or Texas City refinery explosion in 2005. To new engineering graduates Bhopal is ancient history, and they may have been young children when the Texas City refinery explosion occurred. Therefore, it is possible they have never heard of these events.

6.7.4 Countering Estimating and Forecasting Traps

Knowledge of previous incidents in the industry as well as within an individual company can help overcome the Overconfidence and Recallability traps. A forward-thinking company will try to educate new engineers on case studies from the past. There are many resources for these, including CSB reports, the Process Safety Beacon, and *Incidents That Define Process Safety* (CCPS 2008a).

Ask if risk calculations for alternatives been made with equivalent input data and are regarded with the same value.

Be aware that the final decision may depend on more than just risk tolerability criteria and make sure the correct stakeholders are involved in the final decision.

6.8 GROUPTHINK TRAP

Groups of people can make bad decisions by seeking consensus or conformity to minimize conflict. Sources cite groupthink as underlying historical events such as the attack on Pearl Harbor, the Bay of Pigs Invasion, and the Vietnam War. Arguments have been made that groupthink was involved in the Columbia and Challenger shuttle disasters as well.

6.8.1 Groupthink Example, Flixborough, UK Explosion

In 1974, a crack was detected in the 5th reactor of a 6-reactor train process for converting cyclohexane to cyclohexanone. A Maintenance Engineer recommended a complete plant shutdown for 3 weeks for repairs. The Maintenance Manager, who was temporarily appointed to the role and may have lacked appropriate experience, proposed dismantling Reactor No. 5 and connecting numbers 4 and 6 together by a 500 mm (20 inch) diameter temporary connection. To support the piping, the proposal was to use a structure made from conventional construction industry scaffolding.

The temporary connection was not adequate for the forces and temperatures involved and failed, releasing 30 metric tons of cyclohexane in 30 seconds. Of the 28 employees killed, 18 were in the control room. The loss of life could have been far higher if the incident had occurred on a weekday, and not on a Saturday, when the number of day employees on site was low. The whole plant was destroyed. The neighboring housing was devastated. The fire lasted over three days with 40,000 m^2 (10 acres) affected (CCPS, 2008).

The team making the decision was not qualified to do so. The team ignored the advice of the Maintenance Engineer and conformed to the opinion of an unqualified Maintenance Manager.

6.8.2 Countering the Groupthink Trap

Ways to avoid groupthink include (Markman 2015, Govindarajan and Terwilliger 2012, and Hansen, 2103):

- Form a diverse team
- Make the decision as data based as possible
- Be alert for "tweaking" of results to get to a preconceived answer
- Ensure one person, the client, makes the final decision, not the group

- Ask team members to look at the problem as a whole, not just from their department's point of view
- Meet in subgroups, without the client, to evaluate alternatives (to avoid a client's influence)
  - o To achieve even more diversity, after alternatives have been narrowed down, have the subgroups exchange the alternative and do a second evaluation

6.9 SUMMARY

The best way to counter these decision traps is to be aware of them. The *Estimating and Forecasting* traps, especially, are particularly dangerous for decision makers who need to deal with high-consequence/low-frequency events. Traps can work in combination. For example, the Anchoring, Status-Quo, and Overconfidence traps can work hand in hand to blind decision makers to the likelihood of a high consequence event. The Confirming Evidence and Groupthink traps can also work in combination. Risk decision makers can consciously question themselves about whether they are falling into these traps. For example:

- Anchoring and confirming evidence; ask are you weighing all data and calculations equally? If not, why? Ask subgroups to look at the data separately
- Status-Quo; are you improperly favoring the status quo?
- Framing; can you frame the risk decision in alternate, and neutral, ways? If so, do so and see if that affects your view of the decision
- Sunk costs; do you find that previous expenditures of resources are driving your decision?
- Confirming evidence, estimating and forecasting; have risk calculations for alternatives been made with equivalent input data and are regarded with the same value? Be aware of events from other companies outside your own experience. Re-evaluate your decisions if new data or circumstances change.
- Groupthink; Form diverse teams for moderate or complex decisions, respect the inputs of all team members.

Other ways of countering the decision traps include having an unbiased person such as a corporate process safety person on the decision team to understand when these traps are presented and to develop a corporate risk tolerance criterion and enforcing it.

A company with risk tolerance criteria performed a QRA on a storage and handling system for a highly flammable and toxic gas. The system did not meet the company's risk tolerance criteria. Upgrades were recommended that enabled the system to meet the criteria. The business unit director had previously authorized a large capital investment in a new plant that never operated at design capacity due to market forecast errors. Chastened by this experience, he refused to authorize the investment. The corporate safety director, who was also informed of the study results per corporate policy, brought the matter up with corporate directors, who forced the business director to make the upgrades.

Finally, as a suggestion, consider getting *The Art of Thinking Clearly* (Dobelli 2011). This book describes the above decision traps, and many more decision fallacies, in short (2-3 page), quick to read chapters.

7

INHERENTLY SAFER DESIGN

7.1 INTRODUCTION TO INHERENTLY SAFER DESIGN

In Section 2.3 a simplified concept of risk for a hazard scenario was given as the product of the likelihood and the consequences (i.e., Risk = Consequence x Likelihood). Most of the content of the previous chapters has focused on risk decisions involving methods to reduce the likelihood or mitigate the consequences of a risk scenario and choose between them.

The consequences of a scenario are a function of the hazards of the materials or process conditions: flammability, toxicity, reactivity, process temperature, pressure, etc. A more effective approach to risk reduction is to eliminate or reduce the hazards to begin with, when possible. This permanently lowers the consequence term. This is the goal of inherent safety.

CCPS defines inherent safety as "A concept or an approach to safety that focuses on eliminating or reducing the hazards associated with a set of conditions." Inherently Safer Design (ISD) is defined as "A way of thinking about the design of chemical processes and plants that focuses on the elimination or reduction of hazards, rather than on their management and control."

7.2 INHERENTLY SAFER DESIGN STRATEGIES

There are four strategies for implementing inherently safer designs:

Minimize. Use smaller quantities of hazardous substances (also called intensification). Examples would be:

- use of continuous reactors instead of batch reactors, they will contain less material and be easier to control
- storing lower quantities of hazardous materials
- make and consume a hazardous material in-situ to avoid storage

Substitute. Replace a material with a less hazardous material. Examples include running a reaction in water instead of a flammable solvent, finding alternative chemical routes to a product that do not use or create hazardous materials, or use of a heat transfer medium that is compatible with the process fluid.

Moderate. Use less hazardous conditions, or a less hazardous form of a material. Diluting materials or using refrigeration will reduce the required storage pressure and the initial vapor pressure if released. Examples of this are:

- Use aqueous ammonia instead of anhydrous ammonia or hydrochloric acid instead of anhydrous HCl to moderate the consequences of a release.
- Use refrigeration to moderate the conditions at which a material is handled and reduce the amount of vaporization that occurs if it is released.
- Run processes at moderate conditions, e.g., lower temperature and pressures. Use of more moderate processing conditions frequently provides operability and financial benefits as well.

The history of chemical engineering is replete with examples of how processes have been improved and moderated over time (CCPS 2009, p. 54-58).

Simplify. Design facilities to eliminate complexity, making operating errors less likely, or which are more forgiving of errors (error tolerant design). An example is using a vessel that can contain the maximum pressure from a runaway reaction instead of designing an elaborate emergency relief system.

A design becomes more complex when designers and engineers fail to think about alternative design approaches in the early research and development stage of a process's life cycle. Instead, they try to add flexibility and redundancy to systems, default to addition of controls instead of considering alternatives when making changes, or overload a system with alarms, etc.

Inherently Safer Chemical Processes, 2ⁿᵈ Edition (CCPS 2009) provides a more complete description of inherently safer design with more examples of inherently safer designs, plus tools and guidance on approaches to put inherent safety into practice.

7.3 HIERARCHY OF RISK MANAGEMENT CONTROLS

There are four elements to reducing overall risk; they are, in order of preference:

Inherent. Eliminate the hazard using the approaches described in the previous section.

Passive. Minimize the risk with design features that reduce either the consequence or likelihood without active functioning of any device. Dikes, use of pipe-in-pipe, or other secondary containment, blast walls, and flame arrestors are examples of passive protection. Passive controls do not have to operate, but they must be inspected and maintained. Dikes and other forms of secondary containment can develop leaks.

Active. Reduce the risk using control systems, safety instruments systems, mitigation systems, to detect and respond to process deviations.

Active controls require frequent training, testing and maintenance to maintain their reliability. Refrigeration systems are another example of an active system, although they do not have to detect and respond to a loss of containment (LOC) event, they still must be operating to moderate the LOC effect.

Procedural. Reduce the risk through use of policies, operating procedures, inspection, maintenance, training, and emergency response. Good practice is to test procedural controls also with drills or questioning of operators on how they would respond to a process upset during management walk-arounds and audits.

The astute reader will notice there is overlap in the classification of these controls. Refrigerated storage, for example, was listed as an inherently safer design strategy because it moderates the consequence of a release. Some people consider refrigeration a passive control because it does not have to detect and react to a loss of containment event. Others will consider refrigeration an active control measure because it must be monitored and maintained. A well designed (i.e. insulated) refrigerated storage tank can maintain a reduced temperature for a period of time if the refrigeration system fails, allowing time for responding to and correcting the failure.

The concept of ISD can be applied to all levels in the hierarchy of controls. For example, a written procedure can become complex and hard to follow if not well written. With some effort, procedures can be made simpler through use of concise language, common words instead of "$2 words" (e.g., about instead of approximately, or enough instead of sufficient), step-by-step instructions, a common format throughout the facility, the use of flowcharts or pictures, etc.

Figure 7.1 (Amyotte et al. 2006) shows a path through the risk management process that can be applied to the *Identify the Alternatives* step (Section 3.3.3) of a risk decision The risk decision maker needs to consider alternatives from all levels of the risk management controls to improve the quality of the decision.

7.4 ISD EXAMPLES TO ILLUSTRATE DECISION PROCESS

To illustrate the concepts of ISD, some examples of design alternatives with phosgene systems will be used. Phosgene is a highly toxic gas, once used as a chemical warfare agent. The IDLH concentration is 2 ppm, the ERPG-2 is 0.5 ppm, and its vapor pressure is 22.6 psia at 20 °C (68 °F).

> Vapor Pressure: Vapor pressure or equilibrium vapor pressure is defined as the pressure exerted by a vapor in thermodynamic equilibrium with its condensed phases (solid or liquid) at a given temperature in a closed system. The equilibrium vapor pressure is an indication of a liquid's evaporation rate. It relates to the tendency of particles to escape from the liquid (or a solid).

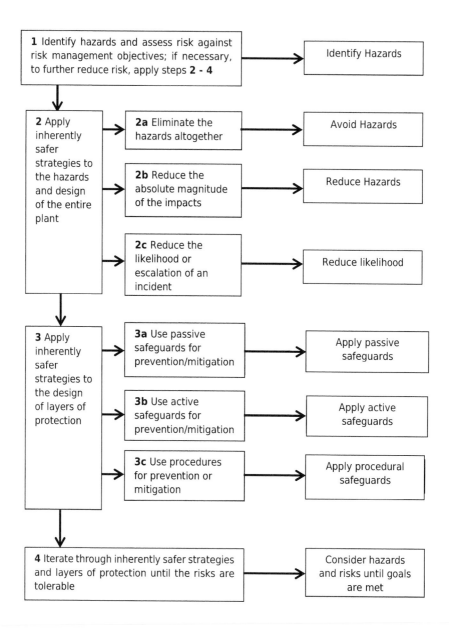

Figure 7.1 A systematic approach to loss prevention and identification of alternatives (adapted from Amyotte et al. 2006)

These examples come from papers published in *Process Safety Progress*:

- Safe Handling of Phosgene in Chemical Processing (Alspach & Bianchi 1984)
- Applying Inherently Safer Concepts to a Phosgene Plant Acquisition (Gowland 1996)
- Toxic Hazard Reduction with Passive Mitigation Systems (Myers, Morgan & Flamberg 1997)

Alspach and Bianchi (1984) describe a phosgene cylinder handling system and list several passive and active safeguards one could include in such a system. These include:

- The phosgene handling system is in a separate building from the rest of the production facility.
- The cylinder connections and associated valves are in an enclosure similar to a glove box to separate them from the operator.
- The enclosure is vented to a scrubber.
- Operators wear supplied air when handling connecting and disconnecting the cylinders.
- Operators are monitored by Closed Circuit TV when in the phosgene building.
- Phosgene is transferred to production via a pipe-within-a-pipe.
- The annulus of the double pipe is vented to a detector or solution of Congo Red that will indicate the presence of phosgene in case of a pipe leak.
- In the production building, the phosgene piping and valves are also totally enclosed.
- Analyzers throughout the phosgene system, in the phosgene building, and above the reactor in which phosgene is used will alarm and close all phosgene valves on detection of phosgene.

This list shows how a combination of inherently safer design principles and active controls are used in the design of a facility handling a toxic material. A few examples will now show how some inherently safer alternatives can be used to reduce risk.

7.4.1 Example with minimization

The Alspach and Bianchi (1986) article described a facility that handled phosgene cylinders. Myers, Morgan & Flamberg (1997) performed consequence analysis calculations on several inherently safer design alternatives with respect to phosgene handling. The base case of this analysis was a pressurized storage tank with 27,000 Kg (30 short tons).

One alternative examined was use of one-ton pressurized cylinders as described by Alspach and Bianchi. This is an application of the inherently safer design concept of minimization. The scenario studied is loss of the entire storage contents in a 10-minute period. Table 7.1 shows the results of the dispersion modeling. The two alternatives do not include other layers of protection, such as use of moderation (see next section) or active safeguards, such as emergency isolation valves.

An organization that has to choose between only these two scenarios will likely have to use ALARP as its risk criteria (Section 2.4.1). This will require adding more inherent, active, and passive safeguards until the risk of one scenario reaches a point where it cannot be reduced any more. Since there are only two alternatives the effort involved makes this an intermediate risk decision.

7.4.2 Example with moderation

Myers, et al. (1997) analyzed the effect of two alternative moderation techniques, alternative dike designs, refrigeration, and containment. Table 7.2 shows the results of these dispersion calculations.

Table 7.3 shows how taking time to identify alternatives (Section 3.3.3) that include inherently safer design options can reduce the consequences of a release and result in better choices for a decision maker.

Identifying hazard scenarios, calculating the consequences and likelihood of the design alternatives may take many man-hours, however, it results in more and better alternatives for the decision makers. In

Table 7.2. Phosgene dispersion distance to ERPG-2. F Stability, 1.5 m/s wind (Myers, et al. 1997).

Alternative	Downwind Distance
27,000 Kg Storage	100 Km (62 miles)
1,000 Kg Storage	34 Km (21 miles)

Table 7.3. Phosgene dispersion distance to ERPG-2 (0.8 ppm). F Stability, 1.5 m/s wind (Myers, et al. 1997).

Alternative	Downwind Distance
1. Pressurized storage, no dike, no enclosure	100 Km (62 miles)
2. Refrigerated storage, low dike wall, concrete dike	14.7 Km (9.1 miles)
3. Refrigerated storage, high dike wall, concrete dike	9.6 Km (5.9 miles)
4. Refrigerated storage, low dike wall, insulated dike	10.7 Km (6.6 miles)
5. Refrigerated storage, high dike wall, insulated dike	5.7 Km (3.5 miles)
6. Refrigerated storage, low dike wall, insulated dike, enclosure	1.5 Km (0.9 miles)
7. Refrigerated storage, high dike wall, insulated dike, enclosure	1 Km (0.6 miles)

addition to the risk calculations, the decision team needs to determine and evaluate other characteristics of the alternatives (Section 3.4 – *Objectives and Outcomes*). Such characteristics can include the amount of risk reduction, active risk controls needed, capital cost, operating cost, corporate reputation, reliability, etc. Therefore, this is an example of a complex decision.

It is plausible that each the proposed alternatives in Table 7.2, when combined with other active, passive, and procedural controls, can meet an organization's risk criteria. Therefore, a weighted scoring technique (Section 3.3.5 – *Making the Decision*) is likely the best tool for this decision. The key to this decision is to identify all objectives and establish the relative importance of each.

If, for example, protecting the corporate reputation is identified as an important objective, i.e., given a high weight, and the plant is near a highly populated area, option 7, which has the shortest impact distance, could be the highest-ranking alternative. On the other hand, if capital cost has a high weight, and the plant is in a remote area, one of the non-enclosure-based options can be the highest ranking alternative.

7.4.3 Example with simplification

Gowland (1996) provides an example of the application of the inherently safer design principle of simplification. In the mid 1980's Dow acquired plants with phosgene production units and began looking at ways to reduce the risk, especially the off-site risk of the units. One inherently safer design approach Dow tried to implement was minimization in the form of reduction of the phosgene inventory. Distillation was necessary to remove impurities in the phosgene before using it in production. The source of the impurities was traced back to impurities in carbon monoxide, a raw material. Dow negotiated a contract with the supplier to obtain a purer grade of carbon monoxide. This allowed Dow to produce a purer grade of phosgene, which eliminated the need for the distillation step. This is an example of simplification and minimization at the same time.

The risk decision in this case is a simple risk decision, or what is technically called a "no-brainer". The real risk decision was actually made early on when Dow recognized the risk and organized a team to study ways to reduce the risk. The team consisted of people from Dow's engineering, safety and business technical center, and they studied many options to reduce risk. That effort had the characteristics of a complex risk decision.

7.4.3 Other tradeoffs

Other risk decision tradeoffs that can be made involving the use of phosgene include:

Make versus buy. The Dow plants made phosgene, but the examples in the Myer's et al. (1997) study were for phosgene storage. Storage can be eliminated by making phosgene on-site. This removes the transportation risk and possibly the storage risk. The likelihood of releases from operational deviations in production of phosgene increases. Additionally, manufacture of phosgene uses chlorine as a raw material, so the risk of chlorine versus phosgene handling needs to be included in the evaluation. Striking the balance between the make versus buy alternative is definitely a complex decision.

Substitution. Substituting a less toxic material for phosgene is an inherently safer option. Such a decision would seem to be a simple one. Like the decision in the Dow (1996) example, however, the amount of work prior to making that decision, in the form of researching alternative chemistries, may be very complex.

7.5 SUMMARY

Figure 7.1 provided a flowchart of a good aid for identifying loss prevention and control options in the *Identify the Alternatives* step of the decision process. Examples of the application of the use of inherently safer designs, as applied to the storage and handling of a highly toxic material, phosgene were presented. These cases showed the power of inherently safer design in reducing the consequences of the toxic releases.

8

MANAGEMENT OF CHANGE

8.1 INTRODUCTION

Management of change (MOC) is a required element of all process safety management systems regardless of regulations. Managing changes is critically important; many incidents (see Table 8.1 for a just a few examples) have inadequate MOC as a management system "root" cause. Most process safety professionals, if forced to rank the generally recognized process safety management elements in order of importance, would probably rank MOC in the top three. Once the need for a change is identified, the risk decision process should be followed.

In the U.S. the OSHA PSM standard requires that a process for managing change be implemented for any change "to process chemicals, technology, equipment, and procedures; and changes to a facility that affects a covered process" that is not a replacement-in kind (RIK) (OSHA 1992). Many organizations now recognize that changes in personnel, staffing levels, and organizational structure should also be covered by a good MOC system. Appendix A of *Guidelines for Management of Change for Process Safety* (CCPS 2008b) is a table of examples of replacement-in-kind and changes for various classes of change. Figure 8.1 is an example MOC system flowchart from the CCPS MOC guideline book.

Table 8.1. Incidents involving inadequate MOC

Incident	Change	Damage	Fatalities
Flixbourough, UK, 1974	Poorly designed bypass pipe installed around a continuous reactor led to flammable release	Major plant damage	28 fatalities 36 injuries
Piper Alpha Platform, North Sea, 1988	Oil platform modified to handle natural gas as well, led to increase risk to crew quarters	Loss of entire platform	167 fatalities
BP Texas City explosion, 2005	MOC of location of portable building did not identify concerns about location even though it violated existing guidelines	> $1.5 billion	15 fatalities 180 injuries
Williams Olefins explosion, LA, 2013	Block valve installed that could isolate reboiler from pressure relief device with only administrative control of the block valve position	Major plant damage	2 fatalities 167 injuries

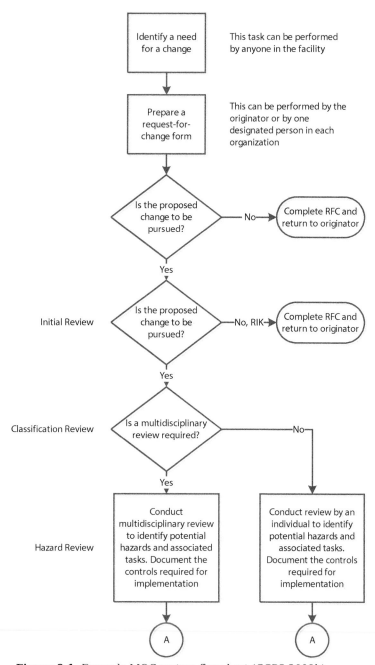

Figure 8.1. Example MOC system flowchart (CCPS 2008b)

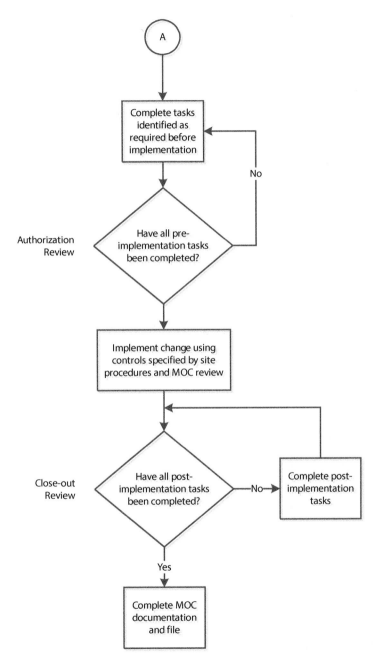

Figure 8.1. Example MOC system flowchart (CCPS 2008b), continued

In addition to an assessment of the risk when changes occur, many additional decisions must be dealt with. These include:

- Process safety documentation (e.g. why the change was made, update of P&IDs, and design basis documents)
- Training (includes why the change is made, the effect on risk; who needs training)
- Operating and maintenance procedures

Before a process change is allowed to be started up, a Pre-Startup Safety Review (PSSR) review is where an acute decision making process may occur. It is at this time when a group may struggle with balancing external demands and process or organizational safety (e.g., in the US, OSHA requires a Pre-Startup Safety Review or PSSR before restarting a process after a change). Understanding the decision between "nice to have" and safety-critical pre-implementation tasks are important.

In addition to *Guidelines for Management of Change for Process Safety* (CCPS 2008b), *Guidelines for Managing Process Safety Risks During Organizational Change* (CCPS 2013) and *Guidelines for Defining Process Safety Competency Requirements* (CCPS 2015a) provide a methodology and a tool, respectively, for analyzing organizational changes.

Temporary changes need to be subject to the MOC process as well as permanent ones. Examples of temporary changes include; bypasses around equipment or interlocks/SIFs, clamps on a leaking pipe, replacing an equipment item with one that is not a RIK until the correct replacement can be obtained, temporarily closing a road in the plant, or a person filling in for someone who has left the organization or is on a leave of absence. When making a temporary change there can be a tendency to accept an increase in risk (either due to an increase in likelihood, consequence, or both) because it should only exist for a short time. Therefore, the authorization of the change has to be at the appropriate level (see Section 8.2).

A good practice is to put a time limit on a temporary change, at which point either; 1) the system must be returned to its original state, 2) the temporary change extended with a new time limit, or 3) the change should be considered for approval as permanent and a new MOC review held with that in mind. The MOC system needs a means to inform the appropriate personnel when the time limit is up.

Although many changes are easy to recognize, e.g. a capacity expansion, revisions to an operating procedure, or a new raw material supplier, some are less obvious.

> A company built a pilot plant for process development of a continuous reactor. Hazard reviews were done for the pilot plant, but not with rigorous methodologies such as What-If or HAZOP.

The research was successful, and the company continued to use the pilot plant for production. This led to running a two or three shift operation instead of one or two shifts, and less technical oversight. The site did not recognize this as a change. After a few months, a spill of a flammable material occurred because a manual valve in the feed system was left open. While the valve was operated only once or twice a day during research, it was operated more often in full production. A review of production records showed it had been operated about 800 times before the error occurred. A full PHA using a formal methodology would have identified the possibility of the error, and a safeguard, such as an interlock to prevent the raw material charge if the valve was open, could have been installed.

8.2 DECISION APPROVAL LEVEL

From a risk decision viewpoint, vulnerability of an MOC system can come from confusion about the level of hazard review and approval needed to make an MOC decision. Usually the risk decision should come well before the MOC which would then be a documentation exercise. However, there are instances where the MOC provides the scope and initial request for further decisions to be made or not. The risk decision documentation is an important input to the MOC and a good MOC process can serve as a second check on the risk evaluation done during the decision process – but normally the MOC process is not used to make the initial decision.

Many companies use the level of capital or expense involved to define the approval level for an MOC. However, changes that involve little or no capital or expense, such as a change in a raw material supplier or the process control system, can still cause a significant increase in risk. Without guidance, there can be a tendency to underestimate the level of hazard review needed for the change, or, conversely, over-analyze a change that poses little risk.

Determination of the level of review and approval requires training of the people who are likely to request the changes on assessing the hazards involved and the effect of the change on safeguards in the process involved. An MOC request form that has a simple checklist to help the change originator assess the types of hazards and the type of reviews needed can assist this assessment. Another way to accomplish this is to designate and train an MOC coordinator to determine the level of review and approval needed.

A shift supervisor or area engineer can approve simple or temporary changes, such as bypassing a process control interlock to finish a batch before repairing a sensor. For some simple changes, it may be possible to identify the need for some common changes in advance and develop specialized forms for them.

Larger organizations may be able to require multiple approvals for all but the simplest changes. The MOC form can have an approval section that resembles the example in Table 8.2.

Appendix C of *Guidelines for Management of Change for Process Safety* (CCPS 2008b) provides examples of MOC forms for simple, moderate, and complex changes. These forms have sections illustrating the approvals needed.

8.3 EXAMPLES OF DECISION PROCESS APPLIED TO CHANGES

8.3.1 Equipment Change

Case Study. In February of 2008, a dust explosion occurred in a belt conveyor enclosure at the Imperial Sugar Company in Wentworth, Georgia. Fourteen people were killed, and 36 injured, some suffering permanent injuries. Secondary dust explosions destroyed the facility. The explosion was the result of an equipment change. The belt conveyor was completely enclosed to prevent contamination of the sugar. The enclosure created space for dust clouds to exceed the Minimum Explosible Concentration (MEC) whereas, before the change, the volume and airflow in the area prevented the dust clouds from exceeding the MEC. The ignition source may have been an overheated bearing (CSB 2009).

There were many contributing factors to this incident, including failure to recognize the hazard from the sugar dust despite many near misses. The focus here, however, is on the decision process. Let's step through a risk decision process as defined in Section 3.3, under the assumption the risk from dust fires explosions was understood at the facility.

Define the problem. The problem was contamination of the sugar, and a goal could be to prevent contamination in a manner that did not increase the risk from dust fires or explosions, within whatever constraints may be in place, e.g., a capital constraint.

Evaluate the baseline risk. Evaluation of the baseline risk shows that the risk of a dust explosion is acceptable (the definition of acceptable

Table 8.2. Example MOC Form Approval Section

Concern	Concern Y/N	Reviewed Y/N	Initials
Occupational Industrial Hygiene			
Process Safety			
Environmental			

depending on the organization's approach to assessing risk). This is due to; the volume of the room and the airflow through the room the conveyor belt is in (which prevents explosible dust clouds from forming), and the facility's housekeeping program and/or fire prevention and suppression systems that prevent or control fires.

Identify the alternatives. Enclosure of the belt conveyor is an alternative, if combined with the appropriate means to prevent or mitigate a dust explosion (control/removal of fugitive dust emissions, venting, suppression, inertion, or explosion containment). Alternatives might be, identify and remove the source of contamination install a cover above the belt but with no sides, or changing to a pneumatic or belt conveyor.

Screen the alternatives. During the screening phase, risk reduction, feasibility, effectiveness, and cost of the alternatives are determined. A semi-quantitative or quantitative risk evaluation method (e.g., a risk matrix or LOPA) is needed to analyze the risk of the alternatives. Some alternatives will probably drop out due to these criteria. Inertion or explosion containment of the enclosure, for example, may be cost prohibitive. Removal of the source of contamination may not be feasible. A cover with no sides may not reduce contamination enough. Pneumatic conveying may be a safer option, if implemented correctly, but more costly to implement. These things cannot be determined, however, if they are not proposed and studied in the first place.

Make the decision. There are several options, each of which is evaluated against a company's risk tolerance criteria, as well as determination of operating and capital cost. The decision team should include representatives from operations, the engineering department, corporate management and corporate health and safety. This is a complex decision.

8.3.2 Procedural Change

Case Study. A power failure caused by a storm delayed plant production for several days. When production resumed, there was urgent demand for product because of the unscheduled downtime. The process in question involved dissolving solid sodium hydroxide (NaOH) in deionized (DI) water, followed by neutralization by mixing with a strong aqueous acid to form the desired product. The neutralization reaction was in a plastic vessel with a loose-fitting top.

The shutdown affected the deionized (DI) water unit, so DI water was not available to dissolve the solid NaOH. The production team decided to add pure undissolved NaOH directly to the vessel into with the aqueous acid already charged. A sudden, severe exothermic reaction occurred, boiling the mixture and causing a gas explosion that cracked the vessel. The operators were sprayed with the chemical solution released. The operators were not wearing the appropriate personal protective equipment (PPE) for a release, and several people were injured. No MOC was performed on the change in the procedure (Ness 2004).

This is an example of a simple decision. A review by someone familiar with the hazard (evolution of heat) involved in an acid-base neutralization, probably a plant engineer or chemist, would have determined that the change was unsafe. Alternatively, operations personnel could have been trained on the reason why the raw material order of addition was the way it was.

8.3.3 Process Parameter Change

Case Study. The Synthron LLC Company received an order for an additive that was 12% more than the size of a normal batch. Plant managers decided to scale up the normal batch to meet the order rather than run two batches. The process involved polymerization of an acrylic monomer in a flammable solvent. A monomer heel and solvent were charged to the reactor and heated. Then monomer and initiator were gradually charged to the reactor. A reflux condenser removed heat from the exothermic reaction.

The 12% scale up was accomplished by adding all of the additional monomer needed to the monomer heel and reducing the solvent charge to allow the increase in the monomer charge. The rate of energy release in the reactor more than doubled, exceeding the cooling capacity of the reactor condenser and leading to a runaway reaction. One person was killed and 14 injured (CSB 2007).

The effect of this change was not effectively reviewed. The fundamental cause of the incident was a lack of understanding of reactive chemical hazards (see Section 4.1.3) by the facilities management and personnel; however, this is a good example of what an intermediate risk decision could be. So once again, let's assume the reactive chemical hazards were understood at the facility and step through the risk decision process as defined in Section 3.3.

Define the problem. The problem definition is how to fill the new order in the most efficient way that maintains the safety of the process.

Evaluate the baseline risk. This was not done at Synthron, but in an organization with awareness of reactive chemical hazards, the process would already have been reviewed by an appropriate methodology. This would likely be a What-If review or a HAZOP (see Section 2.2.1).

Identify the alternatives. Alternatives could include:

1. Two smaller batches

2. Adding the extra monomer to the heel (the option Synthron used)
3. Adding the extra monomer to the gradual monomer feed
4. Adding cooling water to the reactor jacket in addition to using the reflux condenser

Screen the alternatives. The alternatives can be screened easily by doing the heat balance calculations.

Make the decision. Here's how the options stack up:

- Option 1 is obviously safe (no calculations are necessary).
- Option 2 (the one Synthron used) was proven to be unsafe. A simple heat balance would have revealed this.
- Option 3 is safe if the monomer feed rate is properly controlled.
- Option 4 probably requires a capital expenditure given Synthron's equipment set-up.

Options 1, 3, and 4 appear to be safe options. Option 1 is inefficient. Option 4 will require capital investment and extra controls to ensure the reliability of the cooling water flow. Hence, option 3 might be chosen, as it best complies with the problem definition. Monitoring of the temperature and stopping the feed if it exceeds a predetermined point is a requirement for implementing option 3.

8.3.4 Organizational Change

Case Study During a reorganization, a facility shifted its mechanical integrity testing from an in-house group to a third-party vendor. The vendor later sent data pertaining to hydrogen pipeline thickness readings to the in-house group that had been disbanded because a new communication protocol was not provided to the third-party vendor. Therefore, the operations group never reviewed the information. The vendor had identified two pipe elbows that had thinned to a point where they were unsafe. The elbows catastrophically failed after the unit started up, leading to a large fire and pressure wave explosion that damaged the control building, and surrounding process equipment. No loss of life occurred. An MOC was not done for using the third-party vendor instead of the in-house group, which was normally done for equipment, process, or material changes.

Define the problem. After the reorganization, what group will take over management of the third-party vendor and the assessment of mechanical integrity testing results from the vendor?

Identify the alternatives. Alternatives could include:

1. Maintain a position in the facility (i.e., in production or plant maintenance) for an internal subject matter expert to perform this function.
2. Transfer the function to another organization such a corporate level engineering department.

Screen the alternatives and make the decision. This is an example of an intermediate decision. There is a linkage to future decisions, i.e., who will be making future decisions based on the testing results received from the vendor? Similar reorganizations may be planned for other plants that can affect the decision. A person may already exist somewhere in the organization at a corporate level or someone in each plant may have to be trained.

8.3.5 Raw Material Change

Case Study. On April 23, 2012, an explosion occurred in the Lakeland Mills Ltd. sawmill. The explosion travelled east to west through the operating level of the sawmill and collapsed the lunchroom walls onto workers assembled there. Workers in the basement-level millwright's lunchroom were blown out through the south wall by the force of the explosion. The sawmill caught fire as a result of the explosion and was destroyed. Two workers died, and 22 others were injured (WorkSafeBC 2012).

There were many factors involved in this explosion including; ineffective dust control measures, inadequate maintenance of the sawmill gear system (leading to overheating and ignition), the use of open conveyors, and inadequate housekeeping. Also, one mitigation system, a misting system on a debarker, was not operating due to the winter weather causing the piping to freeze.

One other significant factor was a change in the wood processed by mills in British Columbia (BC). In 2005, the Mountain Pine Beetle infected the Lodgepole pine forests in BC. The trees killed by the infestation were allowed to be harvested and processed before the wood lost its value. Studies done when the infestation began found that the wood killed by the beetle was drier than the green wood and, as a result, produced fines with a smaller average particle size, as well as up to 4 times the level of fines than the green wood. The Lakeland sawmill was designed to handle a waste wood and dust volume of 7%. At the time of the explosion, the mill had seen waste wood and dust levels of 12 – 36% (WorkSafeBC 2012, p. 78). The drier, finer wood dust significantly increased the likelihood and consequence of a dust explosion.

As noted, there were many factors involved with the explosion. For this example, we will again assume that the sawmill had equipment and management systems in place that addresses the factors noted in the WorkSafeBC report, and that the change in the wood supply was recognized as a factor to address.

Define the problem. The problem could be defined as how to safely process wood that had been infested with the beetles.

Identify the alternatives. These could include:

1. Do not process this wood.
2. Process the more hazardous wood at reduced rates and/or with administrative controls to manage the hazard (e.g., more frequent cleaning, running only in warm months).
3. Modify plant equipment to handle the more hazardous wood.

Option 2 may entail several separate options, with different combinations of engineering and administrative controls.

Screen the alternatives and make the decision. This is an example of a complex decision. There is both a business and engineering element to the decision; How long can the sawmill expect to have to process the dead wood? Use of the full complement of engineering design and risk analysis tools can be used to screen the alternatives and compare them to the organization's risk criteria. Representatives from the business and the corporate organization should be involved.

8.3.6 Vendor Change

Case study. A plant used 20% Oleum in certain chemical processes. These processes were run in glass-lined equipment. At one point, the Oleum supplier was changed. After several months, vessel inspections revealed that the glass lining had become etched. It was discovered that the new supplier, in addition to making sulfuric acid and oleum, also made hydrogen fluoride. Trace amounts of the HF were in the Oleum, causing the etching.

Define the problem. The problem could be defined as qualifying the new Oleum supplier. A broader problem statement would be to identify the best (or better) overall supplier for Oleum.

Identify the alternatives. In this case, the alternatives are simply the different suppliers.

Identify objectives. This is an example of how important identifying the objectives of the decision. For a raw material vendor change, cost may initially seem to be the only, or at least major concern. Other concerns that can be considered include raw material quality and vendor reliability. In this case, however, the quality problem of HF contamination would not be revealed in a short-term test unless the company knew what other products and processes were run by the supplier. This required identifying the potential for cross contamination at the supplier's plant and auditing their processes for preventing that, possibly with a visit to the plant.

8.4 SUMMARY

Management of change is a critical element of process safety management. Regulatory bodies all over the world recognize this. When the need for a change is identified, the risk decision process should be followed. The depth of review and the approval required is a function of the complexity of the decision, as outlined in Section 5.1.1 - *Types of Decisions*.

Examples of simple, intermediate, and complex decisions involving management of change are presented. Applications of the decision process for changes from real case studies are presented. The risk decision approach is outlined for each case study.

A risk-based decision process enables organizations to make risk decisions consistently and effectively, considering different alternatives and involving the right stakeholders with the use of appropriate methodologies. Changes in facility, organization, process needs to be evaluated critically to ensure risks associated with such changes are understood and risk-based decisions are made.

9

USING LOPA AND RISK MATRICES IN RISK DECISIONS

9.1 INTRODUCTION

Risk matrices and LOPA are order of magnitude risk analysis tools that can be used for some risk decisions. As noted in Section 7.2 – *Risk Evaluation*, Bridges and Dowell (2016) state that 99+% of risk decisions can be made with qualitative and semi-quantitative methods. One can infer these would likely be qualitative risk matrices, semi-quantitative risk matrices, and/or LOPA. In a typical hazard identification risk analysis (HIRA) process, a risk matrix may be used during the hazard identification phase (a PHA), and, if necessary, the results carried over to a LOPA for the risk analysis phase.

9.2 RISK MATRICES

As mentioned in Section 2.3.3, a risk matrix is a two-dimensional chart where the two scales on the matrix describe levels of consequence and frequency. Risk matrices are usually applied during the *Risk Identification* step.

The frequency and consequence ratings set the position for the risk in the matrix. The team can then make recommendations as necessary to lower the risk level to a pre-established level based on the organization's risk tolerance (Section 2.4.1). In constructing a quantitative risk matrix, it is important to note that consequences and frequencies are usually assessed based on orders of magnitude. The risk scores that are obtained when combining consequences and frequencies are additive due to the log nature of individual values. A risk matrix has several advantages as a risk decision tool:

- Risk matrices deal with one scenario at a time and are simple.
- A PHA team can be quickly trained in the use of a risk matrix.
- Risk reduction is demonstrated visually.
- The risk tolerance values (and/or approval requirements) of an organization can be built into the matrix.
- Risk decisions can be made in a consistent and reproducible manner throughout the organization.
- The frequency and consequence categories are usually order of magnitude so the matrix can be used in conjunction with LOPA.

The disadvantages of a risk matrix are:

- PHA and decision-making teams can over or underestimate the consequences of the scenario.
- PHA and decision-making teams may have trouble dealing with or believing very low frequencies outside their everyday experience.
- The risk reduction recommendations developed by a team may not be truly independent, leading to overestimating the risk reduction (a team leader properly trained in the technique can help prevent this).
- The order of magnitude nature of the risk matrix can lead to conservative results.
- Risk matrices are not useful for complex decisions or determining the aggregated risk acceptability of an entire process unit or facility.

9.2.1 Risk Matrix Format

Qualitative risk matrix. An example of a qualitative risk matrix was shown in Section 2.3.3 – *Risk Estimation* and is repeated in Figure 9.1. Tables 9.1 and 9.2 define the qualitative consequence and frequency categories. Table 9.3 defines the risk ranking and response categories of the qualitative risk matrix.

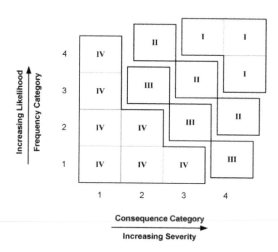

Figure 9.1. Qualitative Risk Matrix example

Table 9. 1. Example Consequence Categories – Qualitative Risk Matrix

Category	Description
1	No injury or health effects
2	Minor to moderate injury or health effects
3	Moderate to severe injury or health effects
4	Permanently disabling injury or fatality

Table 9.2. Example Frequency Categories – Qualitative Risk Matrix

Category	Description
1	Not expected to occur during life of process/facility
2	May occur once during life of process/facility
3	May occur several times during life of process/facility
4	Expected to occur more than once in a year

Table 9.3. Example Risk Ranking/Response Categories – Qualitative Risk Matrix

Risk Level	Description	Required Response
I	Unacceptable	Immediate mitigation or termination of activity
II	High	Mitigation within 6 months
III	Moderate	Mitigation within 12 months
IV	Acceptable As Is	No mitigation required

Quantitative Risk Matrix. An example of a semi-quantitative risk matrix is shown in Figure 9.2. Tables 9.4 and 9.5 define the consequence and frequency categories. In Figure 9.2, risk increases toward upper right (dark), where prompt risk-reduction actions are typically required. Risk decreases toward lower left (white), where risk becomes broadly acceptable and no action is typically required. The intermediate range may require closer analysis, and near-term risk reduction would generally be warranted unless risk is shown to be as low as reasonably practicable.

		Scenario Risk Magnitudes				
Scenario Frequency magnitude	0	1	2	3	4	5
	-1	0	1	2	3	4
	-2	-1	0	1	2	3
	-3	-2	-1	0	1	2
	-4	-3	-2	-1	0	1
	-5	-4	-3	-2	-1	0
		1	2	3	4	5
		Impact Magnitude				

Figure 9.2. Example of a Semi-Quantitative Risk Matrix (CCPS, 2009)

Table 9.4 Example of EHS impact categories and magnitudes used in hazard evaluations for use with Figure 9.2.

	Impact magnitude				
Impact category	1	2	3	4	5
On-site (worker) health effects	Recordable injury	Lost-time injury	Multiple or severe injuries	Permanent health effects	Fatalities
Off-site (public) effects	Odor; exposure below limits	Exposure above limits	Injury	Hospitalization or multiple injuries	Severe injuries or permanent effects
Environmental impacts	Reportable release	Localized and short-term effects	Intermediate effects	Widespread or long-term effects	Widespread and long-term effects
Accountability; attention, concern, response	Plant	Division; regulators	Corporate; neighborhood	Local/state	State/national

Table 9.5 Example initiating cause frequency scale (order-of-magnitude basis) for use with Figure 9.2.

Magnitude 10^x/yr.	Equivalent cause likelihood	Comparison with experience
0	Once a year	Unpredictable as to when it will occur, but within realm of most employees' experience
-1	1 in 10 (10% likelihood) per yr. of operation	Outside of some employees' experience; within realm of process' experience
-2	1 in 100 (1% likelihood) per yr. of operation	Outside of almost all employees' experience; within realm of plant-wide experience
-3	1 in 1,000 per yr. of operation	Outside of almost all process experience; may be within realm of companywide experience
-4	1 in 10,000 per yr. of operation	Outside of most companies' experience; within realm of industry-wide experience
-5	1 in 100,000 per yr. of operation	May be outside the realm of industry-wide experience, except for common types of facilities and operations

9.3 LAYER OF PROTECTION ANALYSIS

Layer of Protection Analysis (LOPA) is a widely used form of semi-quantitative risk analysis. The following items are used in a LOPA:

- *Initiating Event.* The minimum combination of failures or errors necessary to start the chain of events leading to the consequence of concern.
- *Consequence.* The result of the scenario. Examples are: release of toxic or flammable material to the atmosphere, vessel overpressure and rupture, or one or more fatalities due to those consequences.
- *Enabling Conditions.* This is a condition that must exist, e.g. an operating phase such as recycle mode or feed step, which must exist for the accident to occur.
- *Conditional Modifiers.* One or more probabilities included in scenario risk calculations, generally when risk criteria endpoints are expressed in impact terms (e.g., fatalities) instead of in primary loss event terms (e.g., release, vessel rupture). Examples include: probability of ignition, probability of explosion given ignition, or probability of injury or fatality.

- ***Independent Protection Layer.*** An IPL is safeguard that is independent of the event that initiated the accident sequence and independent of any other IPLs identified for the accident sequence during the LOPA. The IPL concept is important and described in more detail in Section 9.3.1.

LOPA typically uses order of magnitude categories for the initiating event frequency, consequence, and the probability of the occurrence of enabling conditions, conditional modifiers, and failure of independent protection layers (IPLs), to analyze and assess the risk of one or more scenarios. The final scenario frequency is the product of the initiating event, the enabling conditions, the conditional modifiers, and the IPLs. Figure 9.3 is an example LOPA table format.

Once LOPA has been applied to yield order-of-magnitude risk estimates for a group of scenarios or consequences, risk decisions can be made. This evaluation is normally in relation to an organization's risk tolerance criteria, see section 2.4.1. The advantages of LOPA as a risk decision tool are:

- A LOPA can be performed more quickly than more complex risk decision tools, sometimes in conjunction with a PHA.
- Scenario identification may be improved through the requirement to identify scenario cause-consequence pairs.
- Rules defining independence of protection layers simplify the need to solve complex situations where safeguards are not fully independent. The risk from different units or plants can be compared as the same method is being used.
- The requirement to determine if protection layers are truly independent may identify operations that were previously thought to have enough safeguards but do not.
- Organizations can develop calculation templates, usually spreadsheets, with built in IPLs and failure probabilities to ensure reproducibility.

Some disadvantages of LOPA are:

- More time and training are needed to properly conduct a LOPA than to use a risk matrix.
- LOPA is an order of magnitude technique, so the results may be conservative.
- LOPA is normally not useful for complex decisions or determining the combined or aggregated risk acceptability associated with an entire unit or facility.

Table 9.6. Summary LOPA Sheet (CCPS 2001)

Scenario Number:	Equipment Number:		Scenario Title:		
Date:	Description			Probability	Frequency (per year)
Consequence Description Category					
Risk Tolerance Criteria (Category or Frequency)					
Initiating Event (typically a frequency)					
Enabling Event or Condition					
Conditional Modifiers (if applicable)					
	Probability of Ignition				
	Probability of Personnel in the Affected Area				
	Probability of Fatal Injury				
	Others				
Frequency of Unmitigated Consequences					
Independent Protection Layers					
BPCS					
Human Intervention					
SIF					
Pressure Relief Device					
Others (must justify)					
Safeguards (non-IPLSs)					
Total PFD for all IPLs	Note: Including added IPL				
Frequency of Mitigated Consequence					
Risk Tolerance Criteria Met? (Yes/No)					
Actions Required to Meet Risk Tolerance Criteria					
Notes					
References (links to hazard review, PFD, P&ID, etc:					
LOPA Analyst and team members:					

Both risk matrices and LOPA:

- analyze one risk (or cause-consequence) scenario at a time
- require some definition of tolerable risk
- require training for the individual or team doing them

A difference between risk matrices and LOPA is that a risk matrix allows one to pictorially demonstrate reductions in both frequency and consequence. LOPA focuses on the frequency portion of the risk equation. Therefore, solutions that reduce the consequence, such as ISD solutions, need to be identified before a LOPA and analyzed as separate scenarios.

Some risk criteria are needed for a risk matrix or LOPA to determine when you have enough layers of protection. LOPA lends itself to more to semi-quantitative risk criteria. With a Risk Matrix, a qualitative criterion such as, "4 IPLs are required to prevent one on-site fatality", can be used.

Once an organization has defined tolerable risk for individual risk scenarios, then risk matrices and/or LOPA can be used for evaluating the qualitative or semi-quantitative risk of each scenario.

If the risk tolerance criteria is defined for a process unit or plant, LOPA may not be the best tool to use, or may need extra guidance to account for aggregate risk, such as establishing a lower risk tolerance of each scenario than the overall unit process unit or plant risk criteria. Another approach may be to enhance the LOPA with better values for initiating event frequencies and PFDs of ILs as described in section 2.3.3. If an organization uses ALARP as its risk objective, risk matrices and LOPA can help a plant (or company) decide if they have reached ALARP but are not sufficient by themselves.

9.3.1 Independent Protection Layers

The concept of an IPL is vital to doing risk estimation via a matrix or LOPA, so the following requirements need to be kept in mind when doing the risk estimation.

In *Guidelines for Initiating Events and Independent Protection Layers in Layer of Protection Analysis* (CCPS 2015), the core attributes of an Independent Protection Layer are listed as:

- *Independence.* The performance of the IPL is not affected by the initiating event or the failure of any other IPL.
- *Functionality.* The IPL must function in a way that prevents of mitigates the consequence of the scenario.
- *Integrity.* The measure of the IPLs capability to achieve the specified risk reduction.
- *Reliability.* The IPL must operate as required, when required, does not experience frequent out-of-service periods, and does not act with out cause.

- *Auditability.* The organization must be able to inspect procedures, records, previous validation assessments to ensure the IPL conforms to expectations.
- *Access Security.* The IPL needs physical and/or administrative controls to reduce the chances of unauthorized system changes.
- *Management of Change.* Changes such as bypassing for testing or temporarily operating with an impaired IPL are reviewed to ensure compensating measures are in place.

It is tempting to count all the safeguards identified in a PHA as IPLs, but those safeguards may not be fully independent, auditable, or deliver the risk reduction needed or an IPL. When using a risk matrix, an IPL needs be able to provide at least one order of magnitude risk reduction. Most organizations have the same requirement for LOPAs.

The reader is referred to *Layer of Protection Analysis: Simplified Process Risk Assessment* (CCPS 2001), *Guidelines for Initiating Events and Independent Protection Layers* (CCPS 2014), and *Guidelines for Safety and Reliable Instrumented Protection Systems* (CCPS 2007b) for more complete information about IPLs.

9.3.2 LOPA Format

A format for performing a LOPA is presented in *Layer of Protection Analysis* (2001) and repeated here in Table 9.6. Another format commonly used is to have the items in the first column put in the top row of a table and the frequency and probabilities in a row beneath them. Either format lends itself to using a spreadsheet to automate the calculations. (Many commercially available PHA documentation programs can automatically convert a cause-consequence scenario from a HAZOP to a LOPA.)

9.4 PHOSGENE HANDLING PROCESS FOR RISK DECISION EXAMPLE

9.4.1 Description.

This example will study a phosgene cylinder handling system as described in the CSB report for the DuPont plant incident (see the **Case Study**). Phosgene is highly toxic, it's ERPG-3 is 1.5 ppm and ERPG-2 is 0.5 ppm.

Phosgene is fed to the process from one of two one-ton cylinders. A 70 psig nitrogen blanket nitrogen pressurizes the cylinder in use and transfers phosgene to the process in which it is used. A ¼-inch diameter, 48-inch long, Polytetrafluoroethylene (PTFE) lined, stainless steel braided flex hose connects the cylinder to the feed line.

Case Study. Fatality at DuPont plant in Belle, WV (CSB 2011). An operator walked into the phosgene cylinder storage area in the Small Lots

Manufacturing (SLM) unit and was sprayed in the face and upper torso with phosgene when a flexible hose ruptured. The worker called for assistance and coworkers immediately went to his aid. His personal dosimeter indicated that he had been exposed to a significant dose of phosgene. He was taken to the hospital. About 3 hours after arriving at the hospital his condition deteriorated, and he died the following night. Two other workers were exposed. The phosgene shed was covered, had walls on two sides, and was open on the other two sides. A sensor at the plant fence line detected a phosgene concentration of 0.27 ppm.

At the time of the incident, the flex hose had not been replaced per the maintenance schedule, and problems with the feeds caused the staff to switch feeds between two phosgene cylinders repeatedly. During those switches the phosgene line was not purged with nitrogen, as when done during a complete cylinder change over. This left phosgene in the line, which expanded during the day and eventually ruptured the hose. Figure 9.4 is a picture of the cylinder connections from the CSB report (CSB 2011.)

When a cylinder empties, operators have to purge the phosgene hose with nitrogen when a cylinder empties, close the nitrogen and phosgene valves on the used cylinder, and then switch to the other cylinder by opening the nitrogen and phosgene valves on it. Operators wear a

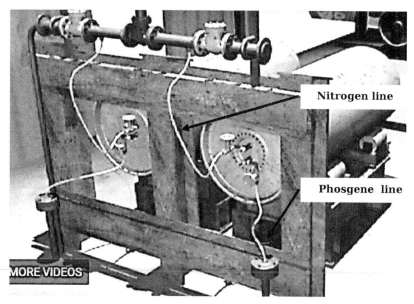

Figure 9.4. Phosgene cylinder hookups at DuPont Belle, WV Plant (courtesy CSB)

supplied air breathing apparatus when performing the switch over. During a walk through to check cylinder weights, only normal PPE is required (gloves, boots and a hood). Several phosgene detectors are located in the shed that alarm in the control room when phosgene levels of 0.05–1 ppm are detected.

Although the actual incident involved both failures to replace the hose and thermal expansion in the flex hose, had the hose not been replaced failure would have eventually occurred due to corrosion alone, even if thermal expansion of the trapped phosgene had not occurred. We already know that flex hose failure of the hose contents can result in on-site fatalities.

9.4.2 Risk Matrix for Phosgene Handling Example

Figure 9.4 shows the risk matrix used for the example process. Tables 9.7 and 9.8 describe the consequence and frequency categories in Figure 9.4.

Scenario Frequency		2	3	4	5	6	7
	1	1	2	3	4	5	6
	2	0	1	2	3	4	5
	3	-1	0	1	2	3	4
	4	-2	-1	0	1	2	3
	5	-3	-2	-1	0	1	2
	6	-4	-3	-2	-1	0	1
	7	-5	-4	-3	-2	-1	0
	8	-6	-5	-4	-3	-2	-1
		2	3	4	5	6	7
		Consequence Magnitude					

Risk increases toward upper right (dark grey, positive numbers), where prompt risk-reduction actions are typically required.

Risk decreases toward lower left (light gray, negative numbers), where risk becomes broadly acceptable and no action is typically required.

Tolerable risk (white, 0). Further risk reduction would generally be desirable unless risk is shown to be as low as reasonably practicable.

Figure 9.4. Risk matrix for example.

Caution. The risk matrix in Figure 9.4.2 was designed to illustrate the logic and technique used in a risk-based decision using Risk Matrices and LOPA. This matrix is not intended as a proposal for specific risk criteria for the process industries.

Examples of risk matrices are presented in *Layer of Protection Analysis: Simplified Risk Assessment* (CCPS 2001), *Guidelines for Hazard Evaluation Procedures, 3rd Ed* (CCPS 2008), and *Guidelines for Developing Quantitative Safety Risk Criteria* (CCPS 2009). The formulation of risk criteria by a company must be accomplished within the context of a variety of considerations (e.g., societal, legal, business, and perhaps, regulatory) that may be unique to the company, and is the subject of *Guidelines for Developing Quantitative Safety Risk Criteria* (CCPS 2009).

Table 9.7 Frequency Categories for Risk Matrix in Figure 9.4.

Magnitude 10 $^{-x}$/yr.	Equivalent cause likelihood per year of operation	Comparison with experience
0	Once a year	Expected but unpredictable as to when it will occur, within realm of most employees'
1	1 in 10 (10% likelihood)	Outside of some employees' experience; within realm of process' experience
2	1 in 100 (1% likelihood)	Outside of almost all employees' experience; within realm of plant-wide experience
3	1 in 1,000	Outside of almost all process experience; may be within realm of companywide experience
4	1 in 10,000	Outside of most companies' experience; within realm of industry-wide experience
5	1 in 100,000	May be outside the realm of industry-wide experience, except for common types of facilities and operations
6	1 in 1,000,000	May be outside the realm of industry-wide experience, except for common types of facilities and operations
7	1 in 10,000,000	Outside the realm of industry-wide experience

Table 9.8 Consequence Categories for Risk Matrix in Figure 9.4.

	Consequence Categories					
Impact category	2	3	4	5	6	7
On-site health effects	Re-cordable injury	Multiple or severe injuries	Hospital-ization or multiple injuries	1 -3 fatalities	4 – 10 fatalities	11 or more fatalities
Off-site (public) effects	Odor	Exposure above ERPG-2	Exposure above ERPG-3	Hospitalization or multiple injuries	1 -3 fatalities	4 or more fatalities

Evaluate the Baseline Risk. Simple dispersion modeling of a release from the phosgene cylinder from a ruptured hose shows that the phosgene ERPG-2 and ERPG-3 concentrations extend well beyond the fenceline, therefore fatalities off-site are possible also. For the sake of the example, we will assume 1-3 fatalities are possible.

Identify the Alternatives. From Section 7.3 – *ISD Examples to Illustrate Decision Aids*, possible risk reduction alternatives that can be examined and their probabilities of failure on demand (PFD) include:

Option A. Switch to a flexible hose with the core, braiding and end fittings maned of Monel®. (This was the type of flexible hose DuPont's own standards recommended for this service.)

Option B. Enclose the phosgene cylinder handling unit and vent the enclosure to a scrubber. Continuously monitor operation of the enclosure ventilation and scrubber in the basic process control system (BPCS).

Option C. Enclose the cylinder connections and associated valves in an enclosure similar to a glove box that is vented to the scrubber. Continuously monitor operation of the enclosure ventilation in the BPCS.

Option D. Require operators wear supplied air whenever in the Phosgene Shed.

Option E. Install analyzers throughout the phosgene building; close all phosgene valves on detection of phosgene and provide a visual and audible alarm to prevent operators from entering the building if a leak is detected.

Option F. Install an automatic valve on the phosgene cylinder outlet that closes if a leak is detected.

Option G. Enclose the entire phosgene cylinder storage and handling system in a building with ventilation and direct outlet air to a scrubber.

Screen the Alternatives. Screening of the alternatives using a risk matrix is demonstrated in Section 9.5, and a LOPA in Section 9.6.

Make the Decision. Decision making based on risk matrix is demonstrated in Section 10.4, and a LOPA in Section 10.5.

9.5 PHOSGENE EXAMPLE DECISION PROCESS USING RISK MATRIX

Evaluate the Baseline Risk. The generic failure rate for rupture of a flexible hose can be between 0.1 - 0.01/year (CCPS 2001 and CCPS 2015).

Note: Although the immediate cause of the hose rupture was thermal expansion, the hose would have failed on its own eventually due to corrosion, so use of this failure rate is justified.

For the hose in question (PTFE/stainless steel) the initiating event rate of 0.1/year will be chosen. This is a Scenario Frequency of 1 in the risk matrix. The consequence is 1-3 fatalities on-site and off-site. These become Consequence Magnitudes of 5 on-site and 6 off-site. The risk categories are **4** on-site and **5** off-site. Figure 9.5 illustrates this.

Screen the Alternatives – Onsite Risk. To reach a tolerable risk a frequency reduction of 4 orders of magnitude is needed. Option A (Monel® hose) reduces the frequency of failure by one order of magnitude. Option B or C (enclosures) reduce the frequency by a second order of magnitude. Option D (full PPE at all times) provides a third order of magnitude reduction. Option E or F the fourth order of magnitude. Figure 9.6 illustrates the risk reduction.

Off-Site Risk. By inspection the need for one more IPL to meet the off-site risk criteria is obvious. Option G will accomplish that. However, the risk analyst or study team needs to be aware that, because Options B or C require interlocks in the Basic Process Control System (BPCS), any additional instrumentation used to monitor the status of the scrubber in Option E needs to be in a separate system, known as the Safety Instrumented System (SIS).

Scenario Frequency						
0	2	3	4	5	6	7
1	1	2	3	**4**	**5**	6
2	0	1	2	3	4	5
3	-1	0	1	2	3	4
4	-2	-1	0	1	2	3
5	-3	-2	-1	0	1	2
6	-4	-3	-2	-1	0	1
7	-5	-4	-3	-2	-1	0
8	-6	-5	-4	-3	-2	-1
	2	**3**	**4**	**5**	**6**	**7**
	Consequence Magnitude					

Figure 9.5. Baseline risk for on-site (light grey) and off-site (dark grey) for phosgene cylinder example.

Scenario Frequency						
0	2	3	4	5	6	7
1	1	2	3	**4**	5	6
2	0	1	2	3	4	5
3	-1	0	1	2	3	4
4	-2	-1	0	1	2	3
5	-3	-2	-1	**0**	1	2
6	-4	-3	-2	-1	0	1
7	-5	-4	-3	-2	-1	0
8	-6	-5	-4	-3	-2	-1
	2	**3**	**4**	**5**	6	**7**
	Consequence Magnitude					

Option A
Option B or C
Option D
Option E or F

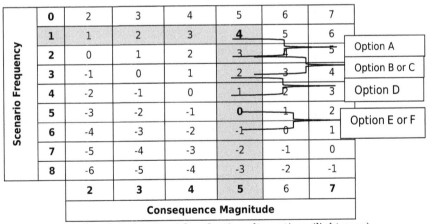

Figure 9.6. On-site risk reduction alternatives (light grey)

9.6 DECISION PROCESS FOR PHOSGENE EXAMPLE USING LOPA

Evaluate the Baseline Risk. Tables 9.9 and 9.10 show the baseline risk for on-site and off-site risk, respectively. In the LOPA, existing safeguards meeting IPL requirements are explicitly considered, so the baseline risk

Table 9.9. On-Site Base Case Risk – Flexible hose failure

Scenario Number: Base Case 1	Scenario Title: Phosgene flexible hose failure with on-site impact	Equipment Number: FH101	
Date:	**Description**	**Proba bility**	**Fre-quency (per year)**
Consequence Description Category	Release of phosgene from full bore rupture of phosgene feed flex hose while operator is in the phosgene shed resulting in 1 – 3 on-site fatalities. (Consequence Category 5).		
Risk Tolerance Criteria (Category or Frequency)	Tolerable risk for on-site fatality = 1/100,000 per year (Frequency Category 5)		1.0E-05
Initiating Event (typically a frequency)	Full bore rupture of flexible hose due to corrosion = 0.1/year. (1)		0.1
Enabling Event or Condition	None	1	
Conditional Modifiers (if applicable)			
	Probability of Ignition	1	
	Probability of Personnel in the Affected Area: Operators in building to check cylinder weights. Total time about ½ - 1 hour/shift.	0.1	
	Probability of Exposure	1	
	Others	1	
Frequency of Unmitigated Consequences			0.01
Independent Protection Layers			
BPCS		1	
Human Intervention	Flex hose is replaced every 3 months	0.1	
SIF		1	
Pressure Relief Device	Not applicable	1	
Others (must justify)		1	
Safeguards (non-IPLSs)		1	
Total PFD for all IPLs		0.1	
Frequency of Mitigated Consequence			1.0E-03
Risk Tolerance Criteria Met? (Yes/No)		No	

Table 9.10. Off-Site Base Case Risk – Flexible hose failure

Scenario Number: Base Case 1	Scenario Title: Phosgene flexible hose failure with off-site impact		Equipment Number: FH101	
Date:	**Description**		**Probability**	**Frequency (per year)**
Consequence Description Category	Release of phosgene from full bore rupture of phosgene feed flex hose resulting in 1 – 3 fatalities off-site. Consequence Category 6.			
Risk Tolerance Criteria (Category or Frequency)	Tolerable risk for off-site fatality = 1/1,000,000 per year (Frequency Category 6)			1.0E-06
Initiating Event (typically a frequency)	Full bore rupture of flexible hose due to corrosion = 0.1/year. (1)			0.1
Enabling Event or Condition	None		1	
Conditional Modifiers (if applicable)				
	Probability of Ignition		1	
	Probability of People in the Affected Area		1	
	Probability of Exposure		1	
	Others		1	
Frequency of Unmitigated Consequences				0.1
Independent Protection Layers				
BPCS			1	
Human Intervention	Flex hose is replaced every 3 months		0.1	
SIF			1	
Pressure Relief Device	Not applicable		1	
Others (must justify)			1	
Safeguards (non-IPLSs)			1	
Total PFD for all IPLs			0.1	
Frequency of Mitigated Consequence				1.0E-02
Risk Tolerance Criteria Met? (Yes/No)			No	

is not exactly the same as for the risk matrix. Neither on-site or meets the organization's risk tolerance criteria.

The probability of failure on demand (PFD) of each option in the LOPA was obtained from *Guidelines for Initiating Events and Independent Protection Layers in Layer of Protection Analysis* (CCPS 2015). The data tables list the design criteria and performance validation methods that need to be met to claim the PFD as an IPL. Some of these options, if used in combination, may need to be Safety Instrumented Functions (SIFs) with their Safety Integrity Level (SIL) determined from the LOPA.

Screen the Alternatives. The alternatives can be screened using LOPA, provided the organization has defined quantitative risk criteria for the consequence.

On-Site Risk: For on-site risk we need at least three orders of magnitude risk reduction. Table 9.11 shows the LOPA evaluation for on-site risk.

A combination of Options B and C enable the risk to be reduced to a frequency one order of magnitude below the tolerable level. By inspection, we can see that Option A can be substituted for Option B and still obtain the same risk reduction. Option A or B alone will enable the process to meet the on-site tolerable risk criteria. However, as further risk reduction is desirable, use of supplied air is a reasonable additional IPL. In a company where ALARP is the requirement, this would be necessary.

Table 9.11. On-Site Mitigated Risk – Flexible hose failure

Scenario Number 1a	Scenario Title: Phosgene flexible hose failure with **on-site** impact	Equipment Number: FH-101	
Date:	Description	Probability	Frequency (per year)
Consequence Description Category	Release of phosgene from full bore rupture of phosgene feed flex hose while operator is in the phosgene shed resulting in 1 – 3 on-site fatality. (Consequence Category 5).		
Risk Tolerance Criteria (Category or Frequency)	Tolerable risk for on-site fatality = 1/100,000 per year (Frequency Category 5)		1.0E-5
Initiating Event (typically a frequency)	Full bore rupture of flexible hose due to corrosion = 0.1/year. (1)		0.1
Enabling Event or Condition	None		
Conditional Modifiers (if applicable)	Probability of Ignition	1	

Table 9.11. On-Site Mitigated Risk – Flexible hose failure (continued)

Scenario Number 1a	Scenario Title: Phosgene flexible hose failure with **on-site** impact	Equipment Number: FH-101	
	Probability of Personnel in the Affected Area: Operators in building to check cylinder weights. Total time about ½ - 1 hour/shift.	0.1	
	Probability of Fatal Injury	1	
	Others		
Frequency of Unmitigated Consequences			0.01
Independent Protection Layers			
BPCS	Enclose the phosgene cylinder handling unit and vent the enclosure to a scrubber. Continuously monitor operation of the enclosure ventilation and scrubber in the basic process control system (BPSC). Note 1.	0.01	
Human Intervention	Flex hose is replaced every 3 months	0.1	
SIF			
Pressure Relief Device			
Others (must justify)	Operators wear supplied air whenever in the Phosgene shed. Note 2.	0.1	
Safeguards (non-IPLs)	Operators are monitored by CCTV when in the Phosgene Shed		
Total PFD for all IPLs	Note: Including added IPLs	0.0001	
Frequency of Mitigated Consequence		1E-6	
Risk Tolerance Criteria Met? (Yes/No)		Yes	
Scenario Number 1a	Scenario Title: Phosgene flexible hose failure with **on-site** impact	Equipment Number: FH-101	
Actions Required to Meet Risk Tolerance Criteria	Options B and C		
Notes	1) Continuous ventilation with automated performance monitoring – Data Table 5.38 (CCPS 2015, p. 236) 2) Personal protective equipment – Data Table 5.49 (CCPS 2015, p. 262)		
References (links to hazard review, PFD, P&ID, etc:			
LOPA Analyst and team members			

<u>Off-Site Risk:</u> Before we even start the off-site risk evaluation, we know that we must meet a stricter risk tolerance criterion, Category 6 instead of 5. We also lose the enabling condition of probability of personnel in the area, and we know PPE will not be an effective IPL. However, an off-site impact may require a larger release, which will probably affect the initiating event frequency. We start off needing four orders of magnitude of risk reduction. Table 9.12 shows the LOPA for off-site risk. Options A or B, and Option D will enable the facility to meet the risk tolerance criteria, if Option D is a SIF with a SIL-2 (PFD = 0.01).

Table 9.12 Off-Site Mitigated Risk – Flexible hose failure

Scenario Number 1b	Scenario Title: Phosgene flexible hose failure with off-site impact	Equipment Number: FH-101	
Date:	**Description**	**Probabil-ity**	**Frequency (per year)**
Consequence Description Category	Release of phosgene from full bore rupture of phosgene feed flex hose resulting in 1 – 3 fatalities off-site. Consequence Category 6		
Risk Tolerance Criteria (Category or Frequency)	Tolerable risk for off-site fatality = 1/1,000,000 per year (Frequency Category 5)		1.0E-6
Initiating Event (typically a frequency)	Full bore rupture of flexible hose due to corrosion = 0.1/year. (1)		0.1
Enabling Event or Condition	None		
Conditional Modifiers (if applicable)			
	Probability of Ignition	1	
	Probability of Personnel in the Affected Area:	1.0	
	Probability of Fatal Injury	1	
	Others		
Frequency of Unmitigated Consequences			0.1
Independent Protection Layers			
BPCS	Enclose the phosgene cylinder handling unit and vent the enclosure to a scrubber. Continuously monitor operation of the enclosure ventilation and scrubber in the basic process control system (BPSC).	0.01	

Table 9.12. Off-Site Mitigated Risk – Flexible hose failure, continued

Scenario Number 1b	Scenario Title: Phosgene flexible hose failure with off-site impact	Equipment Number: FH-101	
Human Intervention	Flex hose is replaced every 3 months	0.1	
SIF	Install analyzers throughout the phosgene building; close all phosgene valves on detection of phosgene and provide a visual and audible alarm to prevent operators from entering the building if a leak is detected. Note 1.	0.01	
Pressure Relief Device			
Others (must justify)			
Safeguards (non-IPLSs)	Operators are monitored by CCTV when in the Phosgene Shed		
Total PFD for all IPLs	Note: Including added IPL	0.00001	
Frequency of Mitigated Consequence			1E-6
Risk Tolerance Criteria Met? (Yes/No)		Yes	
Actions Required to Meet Risk Tolerance Criteria	Options A or B and D		
Notes	The shutdown SIF must be a SIL-2 to meet the risk criteria		

Make the Decision. At this point, with either the risk matrix or the LOPA the risk decision can be made. The facility can install Option A or B, whichever is more cost effective, and Options C and D. This analysis only covered one scenario, however. All possible release scenarios need to be identified and analyzed in the same manner. For simplicity, considerations such as timeline to implement the options, resources needed were not considered in this example, but may be included in the final decision.

The facility has another option before making the decision at this stage, further definition of the consequences if an enclosure vented to a scrubber is installed. If this is done, the off-site consequence, even if the ventilation and scrubber fail, may be a lower category, essentially an inherently safer design option. For example, instead of potential fatalities, the off-site consequence may be exposure to a non-fatal dose. In such a case the consequence category for off-site risk would be 5 instead of 6. If this were the case, the emergency shutdown could become a SIL-1 instead of a SIL-2. However, the organization can opt to stop the analysis at this point and install the selected options.

9.7 SUMMARY

In a real analysis, more scenarios may have to be analyzed. This process, however, is a simple one and there would not be too many more scenarios to consider. Therefore, this is an intermediate decision, and either a risk matrix or LOPA would be satisfactory risk analysis tools for the decision.

10

USING QRA AND SAFETY RISK CRITERIA IN RISK DECISIONS

10.1 INTRODUCTION TO CPQRA

As noted in Chapter 5, QRA is normally applied to complex decisions that involve many risk scenarios, high capital cost and/or severe consequences. A partial QRA, e.g., doing a Fault Tree Analysis (FTA) of a risk scenario with many causes, perhaps combined with a Human Reliability Analysis (HRA), or a detailed consequence analysis of potential loss of containment scenarios combined in an Event Tree, may provide enough information for some decisions. A full QRA, in which the frequencies of multiple risk scenarios are calculated and a consequence analysis of the possible impacts for the scenarios are also calculated, could almost be considered a last step. Figure 10.1 (from CCPS 1999) illustrates the QRA process.

10.1.1 Calculate Frequencies

Sections 2.3.2 – *Frequency*, and 2.3.3 - *Risk Estimation*, covered estimating frequency in a PHA. These estimates can be refined using databases of equipment or system failure rates, human error failure rate estimates or from more analytical methods such as fault and event tree analysis.

Fault Tree Analysis (FTA) is a deductive technique that uses Boolean logic symbols (i.e., AND gates and OR gates) to break down the causes of a *Top Event* into basic equipment failures and human errors (called basic events). Figure 10.2 is an example of a simplified fault tree for the event of a fire in a process area.

Advantages of an FTA: The fault tree can be used to analyze many scenarios at once. Therefore, FTA is useful for complex problems. Also, the results from a fault tree can be more precise than a LOPA, so the result may be less conservative.

Disadvantages of an FTA. An FTA is resource intensive. Creating a fault tree requires a well trained and experienced analyst. Many organizations may need to hire a consultant to do FTA. Creating the fault tree also requires a thorough understanding of the process being studied, requiring a commitment of time from people knowledgeable in the

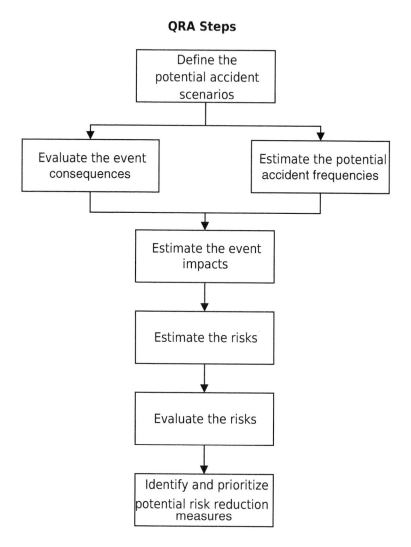

Figure 10.1. QRA process (CCPS 1999)

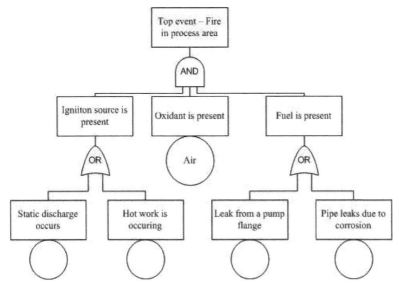

Figure 10.2. Example fault tree for a fire

process and the technology to work with the analyst. This can include operating personnel also. The failure rate data and failure probabilities needed to solve the fault tree may not be readily available, or only generic data may be available, requiring expert judgment to decide what data to use. Finally, since an FTA requires Boolean algebra to solve it, specialized software programs are usually needed to calculate the solution.

Event Tree Analysis (ETA) is a graphical representation of the possible outcomes of an initiating event, given the success or failure of the protection layers. Each outcome thus becomes a risk scenario. Event trees are useful for identifying and illustrating multiple consequences and levels of risk that can occur from a failure in a complex process. By multiplying the initiating event frequency by the probabilities of success or failure of the protection layers, the likelihood of each outcome is determined. Figure 10.3 is an example of a completed event tree.

The advantages of an event tree are similar as those for a fault tree. However, an event tree is simpler to solve than a fault tree, and can be created and solved using a spreadsheet, although software packages are available that can construct and solve an event tree. The event tree is also a good tool for illustrating and explaining the potential outcomes of an event in a simplified manner. In Figure 10.2, the probabilities of each of the failures described in the top row can be estimated using judgement or calculated probabilities.

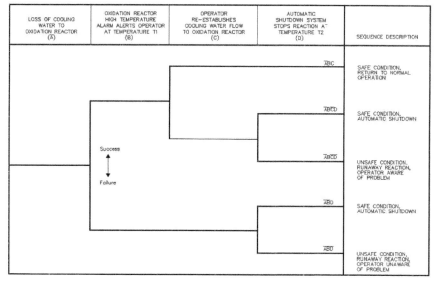

Figure 10.3. Example of an Event Tree (CSB 2008)

Fault trees and event trees are often used in combination. For example, the initiating event in Figure 10.2, *Loss of Cooling Water to Oxidation Reactor*, can be calculated with a fault tree. The probably offailure of the protection layer *Automatic Shutdown System Stops Rea*ction, can also be calculated using FTA.

Bow Tie Diagrams. The combination of FTA and ETA has led to organizations to present risk scenarios graphically in a Bow Tie diagram. An example Bow-tie diagram is shown in Figure 10.4. The left-hand side shows the threats (hazards), causes (initiating events) and barriers (preventive protective layers) of scenarios. The right-hand side shows the barriers (mitigating protection layers) of risk scenarios. Bow Tie diagrams are not calculation methods, instead a Bow Tie diagram is an excellent way of demonstrating and explaining what the risk scenarios of a process are, their layers of protection and potential outcomes to operators and decision makers. An overview of Bow Tie analysis is provided by Pitblado and Weijand (2014) and an in depth study is available in *Bow Ties in Risk Management: A Concept Book for Process Safety* (CCPS 2018).

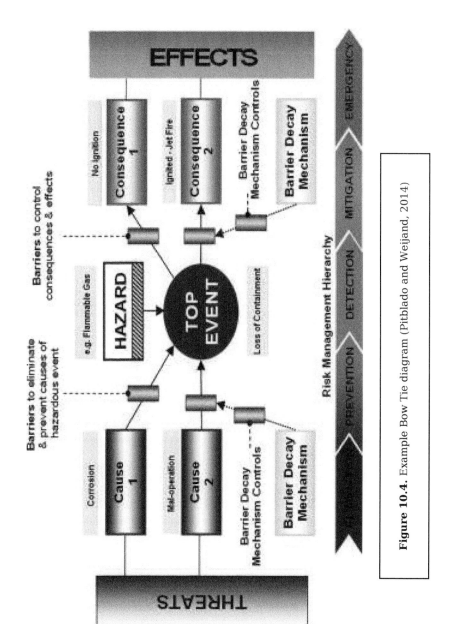

Figure 10.4. Example Bow Tie diagram (Pitblado and Weijand, 2014)

10.1.2 Calculate Consequences

In a QRA the consequences and impacts of a scenario are calculated using a series of models.

Release or source models can calculate the magnitude (the rate and amount) and characteristics of the release (e.g. is it liquid, vapor or a combination) from parameters such as the hole or opening size and orientation, source temperature and pressure, release duration, and so on.

Dispersion models use this release results to calculate the distance to concentrations of concern for the released material, such as specified toxic concentrations or to the lower flammable limit. Dispersion models thus provide an estimate of the area affected by the release.

Consequence models use the results release and dispersion models to calculate the effects of a release. For example, these models can calculate the thermal radiation from a fire, the overpressure from an explosion due to a flammable release, or the distance to defined concentrations of toxic materials due to a toxic release.

Impact models can use the results of the above models to calculate the likelihood of fatalities due to exposure to thermal radiation from a fire, or overpressure from an explosion, or to a dose of a toxic material.

Sources for information on consequence and impact modeling include:

- Guidelines for Consequence Analysis of Chemical Releases (CCPS, 1995)
- Understanding Atmospheric Dispersion of Chemical Releases (CCPS, 1995a)
- Guidelines for Use of Vapor Cloud Dispersion Models (CCPS, 1996)
- Guidelines for Vapor Cloud Explosion, Pressure Vessel Burst, BLEVE and Flash Fire Hazards, 2nd Ed. (CCPS 2010)
- Guidelines for Evaluating Process Plant Buildings for External Explosions, Fires, and Toxic Releases, 2nd Edition (CCPS 2012)

The Center for Chemical Process Safety has spreadsheet-based tools for doing many of the calculations above that are available free of charge. One is Chemical Hazards Engineering Fundamentals (CHEF) and the other is Risk Analysis Screening Tools (RAST). They can be downloaded from the CCPS Resources and Tools page. (https://www.aiche.org/ccps/resources/tools).

The US EPA provides a hazard modeling program known as Aloha®, also free of charge, that can perform dispersion and impact modeling.

10.1.3 Quantitative Risk Analysis (QRA)

As illustrated in Figure 10.1, a QRA uses the frequency, consequence and impact tools described previously to estimate the risk of a process.

The advantage of a QRA is that the contribution of all the identified scenarios to the overall risk can be determined and therefore risk decisions can be prioritized. Then, the effectiveness of risk reduction recommendations can also be analyzed.

The disadvantages of a QRA are:

- QRA is a very resource intensive process, requiring experts in the modeling tools, as well as input from plant operations and process technology.
- The results of a QRA can be more difficult to understand and explain to the decision maker.
- Depending on the skills of the practitioners involved, the results obtained may not be fully reproducible from one study to the next.

Beyond HAZOP and LOPA: Four Different Companies Approaches (Chastain, et al. 2017) describes, as the title suggests, the approach to risk assessment of four companies, Eastman Chemical, Dow Chemical, Celanese, and BASF. In each one, a QRA is the last step after having proceeded through PHAs, LOPAs, FTAs, and ETAs. If required by regulatory authorities a QRA may be done after a PHA.

10.2 SAFETY RISK CRITERIA

10.2.1 Scope of Risk Criteria

Risk criteria can be applied to a single risk scenario, or to an entire unit or facility.

A single scenario risk criterion has the advantage of being simpler to comply with, i.e., a simpler risk analysis tool, such as LOPA, FTA, ETA and/or HRA can be used. If an organization chooses to use a single risk scenario, it may want to make the risk criteria stricter than if applied to an entire unit, to allow for a tolerable risk for an entire unit.

Risk criteria applied to entire unit or facility flips these considerations. More complex risk analysis tools may be required because many risk scenarios are being evaluated. The tolerable risk for each scenario, however, can be estimated by dividing the unit risk criteria by the number of scenarios.

In Section 2.3.3 - *Risk Estimation,* the types of risk criteria were listed as risk indices, individual risk and societal risk.

10.2.2 Individual and Societal Risk

Individual risk measures. Individual risk, sometimes referred to as Geographic Risk, is the risk of a single person exposed to the hazard. Individual risk measures can take several forms:

- *Individual risk contours*; a graphical representation of constant levels of risk around the site of the hazard, in the same manner that topographical contours are shown on a map. Figure 10.5 shows an example of individual risk contours.
- *Maximum individual risk*; the risk to the person exposed to the highest risk in the exposed population, for example to an operator working in the unit.
- *Peak Individual Risk*; the maximum instantaneous risk an individual can be exposed to regardless of how short the duration of the risk exposure is.
- *Average individual risk (exposed population)*; the individual risk averaged over the entire population that is exposed to the specific scenarios, for example all operators in a building.
- *Average individual risk (total population)*; the individual risk averaged over a predetermined population whether or not all people in that population are exposed to the risk.

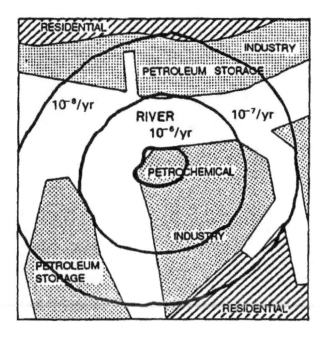

Figure 10.5. Example Individual (or Geographic) Risk Contour

- *Average individual risk (exposed hours/worked hours)*; the individual risk for an activity that is calculated for the duration of the activity or that is averaged over the working day.

Societal risk measures. Societal risk expresses the risk to groups of people who might be affected by events that have the potential to affect large numbers of people. Societal risk is usually expressed as a risk of one or more fatalities per year. They can be applied to on-site and off-site populations. The most common way of displaying societal risk is as a graph known as the Frequency-Number (F-N) curve, where F is the cumulative frequency of all events leading to N or more fatalities. Figure 10.6 shows examples of three F-N curves.

Some countries have established societal risk criteria. In fact, for some companies, a requirement to meet national risk criteria is the only circumstance in which they will perform a full QRA. Figure 10.7 shows the societal risk criteria for The Netherlands. Some organizations have established safety (and environmental and business) risk criteria of their own. One use of QRA is to demonstrate that a process or a facility meets these criteria. This topic is covered in detail in *Guidelines for Developing Quantitative Risk Criteria* (CSB 2009).

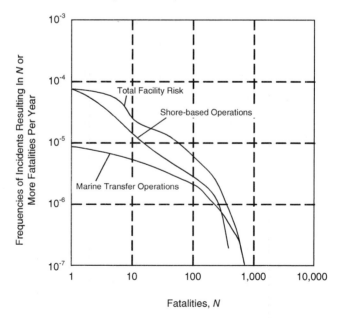

Figure 10.6. Example F-N Curve (CCPS, 2000)

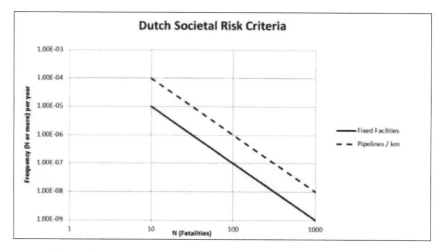

Figure 10.7. Dutch Societal Risk Criteria (Pitblado 2012)

Countries that have risk criteria in some instances include; Abu-Dhabi, Brazil, Belgium, Ireland, The Netherlands, Saudi Arabia and the United Kingdom (Pitblado, et al. 2012).

Beyond HAZOP and LOPA: Four Different Companies Approaches (Chastain, et al. 2017) describes risk criteria in BASF, Dow Chemical, Eastman Chemical and Celanese, and the tools each company uses to do risk analysis. At the time the paper was written:

- BASF uses a semi-quantitative four-by-five risk matrix (Table 10.1).
- Celanese uses a five-by-six risk matrix divided into three regions: intolerable, ALARP, and negligible. Detailed analysis is done using FTA and ETA.
- Dow Chemical has a semi-quantitative risk criteria and uses LOPA, FTA and QRA as necessary. Internal rules are set to provide guidance on when these tools are needed.
- Eastman Chemical uses a combination of the consequences and a risk elevation criterion to determine what tools to use.

Table 10.2 compares the four companies' approaches for complexity and the tools used.

Table 10.1 BASF Risk Matrix

	Severity			
Probability	S1	S2	S3	S4
P0	A	B	D	E
P1	A/B*	B	E	E
P2	B	C	E	F
P3	C	D	F	F
P4	E	F	F	F

*Determined on a case-by-case basis
decision whether A or B is needed.

Probability:
P0 Happened a couple of times (once per year or more often).
P1 Happened once (Approx. once in 10 years).
P2 Almost happened, near miss (Approx. once in 100 years).
P3 Never happened, but is thinkable (Approx. once in 1,000 years).
P4 Reasonably not to be expected (less than once per 10,000 years)

Severity: (Health Effects)
S1 On site: Potential for one or more fatalities.
S2 On site: Potential for one or more serious injuries (irreversible).
S3 On site: Potential for one or more lost time injuries.
S4 On site: Potential for minor injuries or irritation.

Risk Class Minimum Risk Reduction Measures
A Process or design change preferred
B Process or design change, one protective device equivalent to SIL 3
  required (e.g., PSV, SIS)
C Process or design change, one protective device equivalent to SIL 2
  required (e.g., PSV, SIS)
D One monitoring device of high quality with documented testing or
  administrative procedure of high quality
E One monitoring device or administrative procedure
F No technical measures needed

Table 10.2. Comparative complexity and tools used at four companies (Chastain, et al. 2017)

Company	Level of Complexity	Primary Tools*	FTA or Human Reliability Analysis (HRA)	QRA
BASF	Least	Safety Concept + Implementation Checklist (P&ID review, HAZOP, and risk matrix)	Only for low frequency/high severity events	Where regulatory required

Table 10.2. Comparative complexity and tools used at four companies (Chastain, et al. 2017), continued

Company	Level of Complexity	Primary Tools*	FTA or HRA	QRA
Dow	Mid	F&EI, CEI, RC/PHA Questionnaire and LOPA	Special circumstances	For higher risk processes and/or regulatory required
Eastman	Mid	Unit operation specific checklists and LOPA	Special circumstances	Where regulatory required
Celanese	Most	HAZOP, FMEA, checklists and What-If\n\nFor high consequence: FTA, ETA and risk matrix	Frequent	Where regulatory required and to validate simplified approach

*When risk management is needed beyond application of RAGAGEP, regulatory requirements and company specific standards and requirements.

10.2.3 Continual Improvement

"A risk guideline must also drive continuing improvement. Our objective is to improve safety and reduce facility risk in a cost-effective manner. It should not be "good enough" simply to meet a set guideline without an additional requirement for a search for cost effective risk reduction options. The use of a risk range in which risk reduction alternatives must be evaluated for cost effectiveness is one strategy to address the desire for continuing improvement." (Hendershot 1996).

An example of driving continual improvement is BP's major accident risk (MAR) program, described in Section 4.3.1.

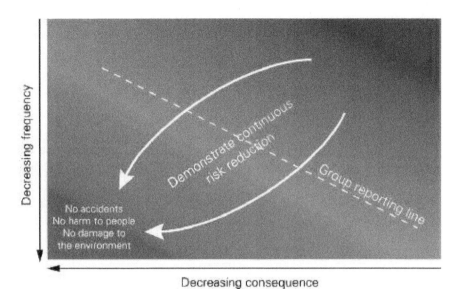

Figure 10.8. BP's group societal risk profile (Considine and Hall, 2009)

10.3 HIGH CONSEQUENCE LOW PROBABILITY (HCLP) EVENTS

Societal tolerance for events that have consequences affecting large numbers of people has decreased over time. Therefore the U.S. public tolerates the deaths of 30,000-40,000 people per year at a rate of 1 – 5 at a time on roads and highways, but the crash of a passenger airliner that kills even a few dozen people at one time receives national headlines.

A High Consequence – Low Probability (HCLP) event could threaten the viability of an organization. Take for example, the explosion at the BP Texas City Refinery on March 23rd, 2005, in which 15 fatalities occurred. About three years later a single fatality occurred at that same facility. The societal response to the first incident was more than 15 greater than the societal response to the second incident.

As a result, F-N curves with slopes of -2 or even -3 are used by some organizations. Thus, the allowable frequency for 10 fatalities may be 1/100th (slope -2) of the allowable frequency for a single fatality or, in other words, a 10-fatality event is 100 times worse than a single fatality event.

A consequence-based approach is another way to deal with HCLP events. For example: a risk-based facility siting study might consider impacts to occupied buildings based on overpressure contours from a ½ inch, 2-inch, and 6-inch leak and then couple the consequence (building damage with potential injuries/fatalities) with the frequency of leaks at those sizes. Based on the frequency of the 6-inch leak, the study might conclude that the even though the building is inside the 6" leak blast contour, the risk is acceptable. A consequence-based approach would be that no occupied buildings are allowed inside the overpressure zones.

HCLP Case Study: The Union Carbide methyl isocyanate (MIC) release in Bhopal, India is an example of an extremely high consequence incident. The exact numbers are in dispute; however, the lower estimates suggest at least 3,000 fatalities, and injuries estimates ranging from tens to hundreds of thousands.

The event occurred when water contaminated a storage tank of MIC, causing an exothermic reaction. There are several theories of exactly what happened. One is that water entered the tank through a common vent line from a source over a hundred meters away. Another is that water was deliberately introduced by a disgruntled employee. There are also theories that water entered from the scrubber over time because the tank was not pressurized or that there was a mix-up in the hose connections for water and nitrogen (CCPS 2008, Macleod, 2014).

Whatever the initial source of the water contamination, there were several failures of other systems that could have mitigated the consequences of the event.

- Pressure gauges and a high temperature alarm which could have warned of the reaction failed.
- A refrigeration system that cooled the liquid MIC was shut down to save money. This could have removed heat from the reaction to prevent or reduce the amount of MIC that boiled up.
- The relief vents of the MIC tank were directed to a scrubber that could have detoxified the MIC, however, the vent gas scrubber was turned off (Figure 10.9).

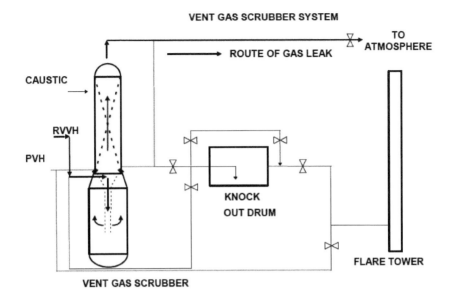

Figure 10.9. Schematic of emergency relief effluent treatment system that included a scrubber and flare tower in series, source AIChE.

- The scrubber was vented to a flare which could have burned the MIC, however, it was disconnected from the process while corroded pipework was being repaired.
- A fixed water curtain designed to absorb MIC vapors did not reach high enough to reach the gas cloud.

A QRA using standard failure probabilities for the layers of protection in place, would likely have shown this to be a low probability event, and perhaps a tolerable risk by a company's risk criteria. The consequence, however, was the worst industrial accident in history. Reasons for this include:

- The community around the plant grew in both size and proximity, exposing large numbers of people to the hazard. An initial QRA would have underestimated the actual consequence if not updated.
- An initial QRA would have overestimated the reliability of the layers of protection. The plant was under severe cost pressures because the product was selling at only 1/3 of the plant's design capacity. Not only were the safety systems shutdown to save money, but maintenance of the plant itself was cut and the plant was in disrepair, and there were cuts in staffing and training. The prioritizing of cost over process safety is evidence of a poor *safety*

culture that was essentially a common cause factor affecting all the layers of protection.

The actual consequence of the incident was so bad that, through spin-offs and mergers, Union Carbide eventually went out of existence.

A good *process safety culture* underlies the reliability of all layers of protection. It requires *maintaining a sense of vulnerability* and *establishing a learning/questioning environment* (Baker Panel Safety Review Panel (Baker Panel 2007)). While it is possible to set criteria for HCLP events by adjusting the slope of the acceptable F-N curve, the most important step in managing the risk of HCLP events is *to identify the events in the hazard identification phase*, and maintaining the protection layers afterwards, not just establishing a criteria for them.

10.4 EXAMPLES

10.4.1 Comparing Design Options: Bromine Handling Facility

A frequent use of individual risk or geographic contours is to compare design options. This is a better use of QRA than reference to fixed criteria as uncertainty in a comparative QRA is much lower than for a QRA done to compare to a fixed criteria. (Note: the calculations necessary to produce geographic risk contours are simpler than the more calculations involved in societal risk.)

An example of this is a QRA of a Bromine unloading, storage and process feed system (Hendershot, et al. 2006). The system consisted of a 750 gallon (2.8 m^3), and a 50 gallon (0.2 m^3) feed tank. Bromine is toxic and has a high vapor pressure. Properties of bromine are listed in Table 10.3. The tanks, and most of the valves and piping were inside a containment building vented to a scrubber and stack. The Bromine was delivered in trucks every few months. Repairs to the storage tank and structural supports were needed to continue operating safely.

Problem definition. The organization operating the system decided to evaluate an inherently safer design (ISD) option as well as repairing the existing system. The problem could be defined as; "Does the risk reduction of the cylinder system justify the expense of removing the existing tanks and installing the cylinder feed process?"

Identify alternatives. The ISD process was to use 120 gallon (0.5 m^3) bromine cylinders instead of the storage and feed tank. The cylinder would be inside the existing containment building, on a scale, and would feed the process directly (no feed tank needed). A second cylinder in a storage area would be stored on-site. All cylinder connections, and all process valves, would be inside the building.

It might seem intuitively obvious that the ISD option has a lower risk, however, the bromine truck unloading only occurred a few times per year, while the cylinder disconnect and connection would occur about twice a month. Therefore the problem definition was to determine if the ISD option provided enough risk reduction to justify the system demolition vs. the repair costs.

Evaluate the baseline risk and screen the alternatives. With the options identified, the next step in the decision process was to evaluate the alternatives, repair vs. replace. This was an existing facility, and the baseline risk (repair) had not been evaluated by a QRA before. This is a common occurrence when looking to replace an existing system. To do the QRA, the risk scenarios for both alternatives had to be identified. Table 10.4 lists the risk scenarios. The frequency of each scenario was calculated.

Table 10.3. Properties of Bromine

Property	Value
Molecular weight	160
Boiling point	59° C
Freezing point	- 7° C
Density at 77° F	25.9 pounds/gallon
Equilibrium vapor concentration at 25° C	285,000 ppm
Permissible exposure limit (PEL) (OSHA)	0.1 ppm
Emergency response planning guideline (ERPG)	
ERPG-1	0.2 ppm
ERPG-2	1 ppm
ERPG-3	5 ppm

Table 10.4 Risk scenarios for bromine unloading and feed system

Truck and storage tank system
Leak in pipe from truck to 750-gallon tank, inside building
Leak from 750-gallon tank
Failure of 750-gallon tank
Leak in pipe from 750-gallon tank to 50-gallon tank
Leak from 50-gallon tank
Failure of 50-gallon tank
Leak from pipe from 50-gallon tank to reactor, inside building
Catastrophic failure of tank truck at normal unloading conditions
Bromine truck overpressurized because of a fire in the unloading area or adjacent building—heat
exposure and boiling of bromine, relief valve fails to operate
Bromine truck overpressurized because of a fire in the unloading area or adjacent building—heat
exposure and boiling of bromine, relief valve operates properly resulting in a discharge, but not a rupture of the truck
Various size leaks in truck shell
Unloading hose failure with rapid response (10 min) by operator to shut off the flow
Unloading hose failure with no response by operator to shut off the flow. The flow continues for 1 h

Table 10.4 Risk scenarios for bromine unloading and feed system, continued

Cylinder system
Unloading hose failure (2 hoses), or hose not properly connected
Cylinder valve leak
Catastrophic failure of the bromine cylinder (including dropping or mishandling)
Failure of pipe, or leak in pipe from cylinder to reactor
Leak directly from cylinder
Leak directly from cylinder, outside building

The risk analysis consequence and risk calculations were done using a risk analysis computer program, in this case SAFETI 6.4 from DNV.

Figure 10.10 shows the resulting individual risk contours for the truck/storage tank and Figure 10.11 for the cylinder system. Table 10.5 provides the results in tabular form. In Option B occupied residences are not affected at all.

Make the decision. The cylinder system provides several orders of magnitude of risk reduction. Assuming that cost of dismantling the existing equipment is not disproportionate the risk reduction, replacement is an easy decision.

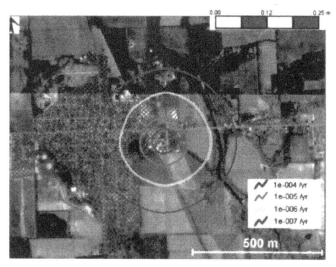

Figure 10.10. Individual risk contours for bromine storage tank system (Hendershot 2006).

Figure 10.11. Individual risk contours for bromine cylinder system (Hendershot 2006).

Table 10.5 Distances to Individual Risk Contours in Figures 10.10 and 10.11

System	Risk Contour, yr-1	Distance, m (ft)
10.2 Storage Tank	10^{-4} (1/10,000)	50 (164)
	10^{-5} (1/100,000)	100 (328)
	10^{-6} (1/1,000,000)	185 (607)
	10^{-7} (1/10,000,000)	280 (919)
10.3 Cylinder Handling	10^{-7} (1/10,000,000)	50 (164)
	10^{-8} (1/100,000,000)	80 (262)

This is an example of an intermediate to complex risk decision. There were only two alternatives, but there were several risk scenarios. If the design was a new facility, there could be more alternatives that could be evaluated, such as the location of the unit with respect to the nearest population or occupied buildings.

10.4.2 Compliance and Continual Improvement: Organic Acid System

An example of using QRA to drive continual improvement is a study done by Eastman Chemical on the process vent systems handling organic acids (Garland 2010). Eastman Chemical has hundreds of vessels using acetic acid, propionic acid, and butyric acid as solvents. These were designed according to the appropriate engineering standards of their time, some dating back to the 1930's. In the U.S., the cutoff between a flammable and combustible liquid is 100 F ° (38 °C). A material like acetic acid, flash point 104 °F (40 °C), is close to this man-made cutoff, so, it is not difficult to envision process conditions that would put an acetic acid containing process stream above its flash point. Eastman decided they wanted to confirm these systems complied with their current risk standards and develop risk reduction alternatives if not.

Problem statement. Eastman wanted to know if the systems met current risk criteria, and if not, should nitrogen inertion be added to any of the vessels and vent systems.

Evaluate baseline risk. To evaluate the baseline risk, flash point and MIE data had to be collected for the acid/water mixtures at various conditions. With the data in hand, a qualitative risk assessment of the vessels containing the organic acids was done to identify those systems with risk scenarios requiring further studies. Decision criteria were developed for grouping types of vessels together and eliminating those systems that were clearly not contributors to the risk (i.e., were not operating in flammable zones.

Identify the alternatives. The alternative was identified in the problem statement, nitrogen inertion.

Screen the alternatives. A QRA was done for the vessels that were not eliminated in the qualitative screening phase. Table 10.2 shows that Eastman's preferred tools are LOPA and FTA/HRA. A key input for the QRA was to estimate ignition sources and probabilities.

Make the decision. The QRA showed that the vessels and vent systems met Eastman's internal risk criteria without nitrogen blanketing.

Continual risk reduction. The combination of qualitative and quantitative risk assessment identified 13 risk reduction strategies that could be implemented enable to comply with the ALARP concept.

This was an example of a complex risk decision. It involved many possible systems and required data collection, qualitative and quantitative risk assessment to make the decision.

10.4.3 Special Case: The Domino Effect

CCPS defines the domino effect as "The triggering of secondary events, such as toxic releases, by a primary event, such as an explosion, such that the result is an increase in consequences or area of an effect zone. Generally only considered when a significant escalation of the original incident results." The HSE document *Development of methods to assess the significance of domino effects from major hazard sites* (HSE 1998) discusses the phenomena in detail. Once identified, domino events can be included in a QRA by adding them as new risk scenarios, then calculating their frequency and impact.

Case Study 1, Onsite Domino Effect, PEMEX Mexico City LPG Terminal. A good example of the domino effect occurred at an LPG terminal in Mexico City in 1984. The terminal had four 1600 m3 LPG spheres, two 2,400 m3 spheres, and 48 horizontal cylindrical storage tanks of various capacities. The cause of the release has never been established, but it was likely either a tank overflow or overpressurization of an LPG line. After about 5-10 minutes the gas cloud drifted to a flare stack. It ignited, causing violent ground shock. A number of ground fires occurred. The degree of congestion in the horizontal storage tank enclosure was such that the fires caused nine explosions and BLEVE's (Figure 10.6). Tanks were thrown off their supports and piping ruptured. The four small spheres were completely destroyed with metal projectiles going as far as 350 meters (1150 ft.) away into public areas. Twelve of horizontal tanks were ejected over 100 meters (330 ft.) from their supports, with the furthest going 1200 meters (3940 ft.). Only four of the horizontal cylindrical tanks survived. Gas that entered buildings inside the terminal and into public housing ignited resulting in explosions. Workers in the plant now tried to deal with the release by taking various actions. Flow to the terminal continuing for one hour after the explosion. This incident resulted in over 600 people killed and about 7,000 injured (Lees 1996 and CCPS 2008a). With appropriate level and pressure controls, the likelihood of the incident happening in the first place could have been reduced. With appropriate remote emergency shutdown controls, better fire protection, and spacing between storage tanks, the extent of the consequences from the initial fire could have been limited. *Guidelines for Siting and Layout of Facilities, 2nd Edition* (CCPS 2018), provides spacing tables for making these informed decisions.

Figure 10.6. PEMEX LPG terminal showing domino effects, note multiple fires

Case Study 2, Near Miss Offsite Domino Effect, First Chemical Nitrotoluene Explosion, 2002. A distillation column was shut down, with a heel of 1,200 gallons (4.5 m³) of nitrotoluene in the bottom. Steam leaked through a valve and heated the nitrotoluene to its decomposition temperature over a period of several days. The column exploded. One piece of the column hit a nitrotoluene storage tank, and another a cooling tower on-site. A fire started at each place. A 6-ton (6,096 kg) piece of the column traveled 1,100 feet (335 m) and landed near a crude oil storage facility (CSB 2003).

Domino effects are most often caused by fires and explosions, however a toxic release can also cause one. For instance, in the phosgene example in Chapter 9, the release could have entered a control room, incapacitating operators and causing loss of control and containment at another process.

As with the high consequence low probability (HCLP) events, it is necessary to identify the potential for domino effects in order to assess and make decisions about risk mitigation measures. *Development of methods to assess the significance of domino effects from major hazard sites* (HSE 1998) provides guidance on this. *The Treatment of Domino Effects in Process Hazard Analysis* (Baybutt 2015) provides a method for identification of domino events by a combination of identifying events that can originate in a unit and by looking for events that can impact the unit. Baybutt then recommends some form of QRA for a detailed analysis.

To identify the potential for domino effects, a consequence analysis of the impact of fires, explosions and toxic releases can be done to determine the extent of:

- the spread of pool fires (direct thermal impact on other equipment items and structural supports)
- thermal radiation from fires
- overpressure from explosions

- projectiles from explosions
- hazardous concentrations from toxic releases

During the analysis phase risk mitigation factors that can be tested include:

- Safety Instrumented Systems (SIS)
- remotely operated shut-off valves
- inventory reduction
- separation of equipment items and units
- dikes
- drainage to safe locations
- fire and or blast walls
- LFL or toxic gas detectors with alarms and/or automatic shutoffs
- automatic fire protection
- automatic or manual ESD to shutdown the whole process
- orienting horizontal bullet storage tanks to avoid pointing at other vulnerable targets

10.5 SUMMARY

Some countries and many companies use risk criteria and QRA in their decision-making processes. The risk criteria and QRA tools and strategies of five companies, BASF, Dow Chemical, Eastman Chemical, Celanese, and BP, were described. The risk criteria employed ranged from semi-quantitative risk matrices, to ALARP, to individual and societal risk criteria. Tools employed included LOPA, FTA, ETA to full QRA.

Two examples of how companies used QRA to make risk decisions were presented. One was an intermediate level decision to choose between two options for upgrading an aging facility. Another was a complex decision that involved hundreds of vessels. The company employed a combination of data acquisition, qualitative risk assessment, followed by QRA to determine if the vessels met the companies risk criteria and identity further risk reduction options to go beyond just meeting the criteria.

Two special cases of risk problems were also described. The High Consequence - Low Probability (HCLP) event and domino events. In each of these cases, the most important step is to identify the events that need to be evaluated. Once identified, the evaluation is straightforward. Maintaining a sense of vulnerability is key to this.

Communication of the results of a QRA up and down the organization requires turning data analysis into "a summary" that all people understand and providing access to the data for informed decision-making.

11
DECISION IMPLEMENTATION

11.1 INTRODUCTION

This chapter discusses implementation of the decision. Making, implementing, and documenting risk decisions will always involve a group of people. Even a quick or simple decision, which might be made by a single person, may require other people to implement.

11.2 IMPLEMENTATION

Unless a decision is to do nothing, i.e., keep the status quo, then, by definition, implementing the decision means a change to the facility, hence a Management of Change (MOC) review is required. The MOC process is described in more detail in Chapter 9. *Guidelines for Management of Change for Process Safety* (CCPS 2008b) describes the MOC process in detail.

11.3 DOCUMENTATION

11.3.1 Importance of a decision document

The members of the decision team may or may not be involved with implementing the decision. The documentation is a guide to those who do. The documentation can be contained in more than one document. For example, the documentation of the application of the risk analysis tools (PHA, Risk Matrices, LOPAs, FTA/ETAs, and QRAs) and/or the documentation of the basis for alternative designs can be separate documents from, or perhaps appendices to, the one describing the decision and how it was made.

Parts or all of the documentation can become part of the process safety information about the process. These documents will contain information about the technical assumptions and basis for the decision.

11.3.2 Writing recommendations.

In addition to the final decision itself, there can be other recommendations that come out of the decision analysis. For example, if a risk analysis tool is used to compare options, the use of the tool may reveal further recommendations for risk reduction within the chosen alternative.

FTA, ETA, QRA, and perhaps enhanced LOPA, tools can not only calculate the overall risk, but also determine the contribution of individual scenarios to the total risk. So, for example, in the case study in Section 11.4.1 – *Comparing Design Options*, the contribution of each scenario in Table 11.4 – *Risk scenarios from bromine handling and feed system*, will be calculated by the QRA software. The decision team, or the decision implementers, may find ways to reduce the overall risk further by addressing some of these scenarios.

Recommendations should be written in enough detail for a different group to implement it. A good practice is to write the recommendation in such a way as to leave the implementers options to address the recommendation. A PHA team should focus on hazard identification as opposed to solutions, and another group of people can take time to evaluate how to best implement the safeguards required. In the same way a decision team should focus on the decision itself.

In the case study in Section 12.2.1, the key decision was which plant to install the process at. It is easy to imagine that while doing an initial design of the retrofit and new plant options for cost estimating purposes, the design teams may have come across design decisions that also had alternative solutions. Unless this subset of decisions has a material effect on the outcome, these options should not be evaluated until later. The decision documentation can record the design options.

11.3.3 Advice of legal counsel.

If the decision is made based on the organization's existing risk criteria, and the risk criteria was established at the corporate level, with guidance or input from the legal department, then there should be no need to bring in legal counsel. Likewise, if the process involved is in a country with a regulatory risk criterion, and it meets those criteria, then there should be no need to bring in legal counsel.

If, however, this is not the case, and a decision team is developing its own criteria, then legal advice may be necessary.

Case study. A manufacturing plant needed to ensure that allowable peak styrene loads to a Publicly Owned Treatment Works (POTW) were not exceeded (Davis and Ness, 1999). This had already happened several times, and the POTW and the city were threatening large fines and a possible plant shutdown if it happened again. As the project began, one alternative had already been defined, collection and analysis of all waste before releasing it to the POTW. The team was given the task of identifying and analyzing other options. Although the organization had tolerable safety risk criteria, no environmental risk criteria was ever established. For the purposes of this analysis the team proposed a tolerable environmental criterion for exceeding the peak styrene loads. The decision criterion was shared with the business and corporate legal department before discussing it with the city. When discussing the criteria with the city, the organization also explained the methodology that would be used. This was:

- Identify applicable release scenarios from previous PHAs
- Identify equipment failure release scenarios using Failure Mode and Effects Analysis (FMEA) on equipment items handling Styrene
- Identify operational release scenarios using a guideword analysis of styrene loading and unloading operations
- Determine the likelihood of releases identified above using LOPA
- Identify risk reduction recommendations to reduce the total frequency to below the proposed criteria

The city accepted the organizations proposal for a tolerable release frequency, in no small part due to the openness with which the organization explained the risk analysis procedures.

11.3.4 Contents of the decision document

Good documentation will:

- Contain the problem definition, explained in more detail if necessary. The problem definition is needed for the readers to understand how the team saw the problem. The team can explain how they came to define the problem in the way they did.
- Present the alternatives considered. The alternatives are needed so the readers understand the options considered. At this point the document can also address options that were not considered and why.
- Present the alternative chosen, and why it was chosen. This should include the decision criteria and decision tools as outlined in Chapter 3.
- Present key assumptions. Readers need this information in case circumstances occur that change or affect these assumptions. They may not be entirely technical assumptions. They can be assumptions about the production levels, surrounding population, etc.
- Explain why the other alternatives were not chosen. What criteria did each unchosen option fall short in?
- Contain key references. This can be PHAs, risk analyses, design documents, cost benefit analysis, etc.

11.3.5 Retention of the decision document

A decision document can contain process safety information (PSI). This is information about the chemicals, technology and equipment in the process. If the decision process document contains new process safety information, it should be added to the existing PSI documentation and kept for the life of the process. PSI should always be kept up to date as changes are made or new information about the process or its hazards arises.

In the U.S. an organization is required to keep PHA and PHA revalidation documentation for the life of the process. Other countries or local authorities may have other requirements. If a PHA of a covered process was done as part of the risk decision process, then that is covered and needs to be kept for the life of the process.

If no legal requirements exist, and there is no process safety information in the document, then an organization can decide on a records retention policy. If a PHA or any risk analysis methods are used, a good practice would be to keep the documentation long enough for one or two rounds of PHA revalidation.

11.4 REVALIDATION

The decision itself may need to be reviewed in the future, depending on circumstances.

11.4.1 Time based

Local regulations may require a revalidation based on time. For example, as stated above, with a US based PSM covered process revalidation of the PHA is required every 5 years. Process safety studies such as LOPAs, FTAs, and QRAs may need to be updated during or after this revalidation. Some companies in the US have chosen to revalidate on a shorter schedule, for example three years, whether a process is covered by the PSM regulation or not.

11.4.2 Situation based

If significant changes in the facility, or the assumptions that went into the analysis occur, an organization may choose to revalidate the analysis at that time. Examples of assumptions that could affect the decision basis (other than a significant facility change) are:

- A significant change to the process operation (a new equipment item, a second production train).
- A change in operating conditions (e.g. use of higher temperatures and pressures).
- A change in on-site risks (i.e., another production unit installed nearby (see 10.4.3 – A Special Case: The Domino Effect) or on-site population
- A change in off-site population (e.g., an occupied building built close to the facility).
- A change in production levels that effects on-line time (e.g. a switch from 5 day/week operation to 7 days/week, or vice versa).
- A change in the number of raw material deliveries (more connections and disconnections).

11.5 SUMMARY

The decision team make-up depends on the nature and complexity of the decision. A representative from the areas of key technical and process inputs should be on the team. Specialists, e.g. cost analysts, process safety representative, can be team members or brought in as resources as needed. For intermediate and complex decisions, an opening meeting is needed for decision team members to get to know each other, define the problem, and identify alternatives. The team should use this time to identify the resources needed to screen alternatives.

Implementation of the decision may be done by people not on the decision team. Documentation of the decision and assumptions needs to be complete and clear. It should include:

- Problem definition
- Alternatives considered
- Alternative chosen
- Why that alternative was chosen
- Key assumptions
- Key references

Implementation of the decision will require following the MOC process. That means possible PHA studies, as well as updating the PSI and operating instructions and training operators.

12

SUMMARY AND LESSONS

12.1 INTRODUCTION

Following the decision process in this book (or any well-defined decision process) will result in better outcomes for most cases. In Chapters 7 through 11 examples of how to apply the decision process were presented. In this chapter, examples of the failure to apply the steps of the decision process are presented with case studies.

As a reminder, the decision process steps are:

3. Define the Problem
4. Evaluate the Baseline Risk
5. Identify the Alternatives
6. Evaluate the Alternatives
7. Make the Decision

12.2 CASE STUDIES IN RISK: DECISION MAKING FAILURES

12.2.1 Failure to Define the Problem.

MFG Chemical, Runaway Reaction and Toxic Release, Georgia, 2004. A runaway chemical reaction occurred at MFG Chemical that released a vapor cloud of allyl alcohol and allyl chloride (a toxic chemical). The vapor cloud drifted towards a residential community and 154 people needed decontamination and treatment for exposure to the cloud. The first emergency responders were told the incident was a spill but there was no mention of a vapor cloud. The first responders drove through the cloud and were affected by the allyl chloride. It was these responders who reported the event as a toxic release. Other emergency responders had to go into the community without the proper equipment in order to evacuate the residents (CSB 2006).

MFG identified three process alternatives. Two alternatives used a gradual feed of one of the reactants, and one added all the reactants at one time and heated the batch. MFG chose the simplest, add all of the reactants at once. This is the least safe way to run a batch exothermic reaction. This event is an example of not defining the problem and objectives. The decision objectives did not include process safety objectives. This is also an example of poor application of the RBPS elements of *Process Safety Competency* (they did not appear to know they had a problem as they did not follow the inherently safer principles at the initial design phase), and *Emergency Management*.

12.2.2 Failure to Establish Baseline Risk and Identify Alternatives

Hickson PharmaChem, Runaway Reaction and Explosion, Ireland, 1993. A decomposition runaway reaction and explosion occurred in a vacuum distillation unit for recovering isopropyl alcohol (IPA) from a mother liquor containing nitro compounds. The explosion caused one major injury, damaged equipment, and destroyed a process building (Birtwistle, 2001).

In the plant, a batch distillation was done to recover IPA and the residue was cooled and disposed of after every batch. In 1991 the plant switched to a multiple step distillation. Three distillations of IPA were run before the residue was removed. The extra distillation steps exposed the residue, containing thermally unstable compounds, to heat for longer periods. As a result of the process change, the concentration of thermally unstable material in the residue increased. In 1993 a power outage occurred resulting in loss of cooling and agitation, as well as loss of the temperature and pressure readings in the still pot. Shortly after the power went out the explosion occurred.

Hickson did not evaluate the *baseline risk (step 2)* of the existing process (single distillation) or evaluate the alternative process (multiple distillations) by doing a *hazard identification and risk analysis* review, a core RBPS element. This is also an example of inadequate application of the RBPS element *management of change*.

Hickson did not *identify alternatives (step 3)* (Section 12.2.3). Rather than keep distillation residue in the batch still, the IPA solution could have been stored in a separate vessel until there was enough to do the equivalent of three batches, thus not exposing residue to prolonged heating. Another, inherently safer, alternative would have been to do a continuous distillation with a thin or wiped film evaporator, minimizing exposure time of residue to heat.

Imperial Sugar, Dust Explosion, Georgia, 2008. This incident was described in Section 8.3.1 – *Equipment Change*. A dust explosion, followed by a series of secondary dust explosions, occurred at the Imperial Sugar refinery in Port Wentworth, Georgia in February 2008 that killed 14 people and injured 36, some permanently. The explosions destroyed the facility. Imperial Sugar was fined $8.7 million. A complete report and a video describing the event are available from the Chemical Safety Board (CSB 2009a).

Not only did Imperial Sugar not *establish the baseline risk (step 2)*, but they were not following relevant standards regarding dust accumulation such as NFPA 654 *Standard for the Prevention of Fire and Explosions from the Manufacturing, Processing, and Handling of Combustible Particulate Solids*. If the base line risk had been evaluated, they would have found that the airflow through the tunnel prevented a buildup of combustible dust above its minimum explosive concentration, and that enclosing the conveyor without that airflow would have resulted in the accumulation of dust inside the enclosure.

Define Alternatives. Rather than entirely enclose the belt conveyor Imperial Sugar could have considered installing a cover without sides over the belt or maintaining an airflow through the enclosure to prevent dust accumulation. Alternatively, the belt conveyor could have been replaced by a screw conveyor (several of these were already used in the plant).

12.2.3 Make the Decision - Failure to consider tradeoffs

Esso Longford Gas Plant, Explosion, Australia, 1998. A major explosion and fire occurred at Esso's Longford gas processing site in Victoria, Australia. Two employees were killed, and eight others injured. The incident caused the destruction of Plant 1 and shutdown of Plants 2 and 3 at the site. This incident reduced gas supplies to the entire state to 5% of normal, causing major business disruption, and a lot of cold showers, for the inhabitants of the province (CCPS 2008a).

A process upset in a set of absorbers eventually caused a series of automatic shutdowns. This caused a loss of flow of a hot "lean oil" stream to a heat exchanger, resulting in a large temperature decrease. (In some parts of the plant temperatures reached -48 °C (-54 °F). The decrease in temperature allowed the metal heat exchanger to become extremely cold and brittle. The exchanger started leaking, and operators eventually restarted flow of the hot lean oil to the heat exchanger to try to reduce the leaks. The resulting thermal stresses caused the heat exchanger to rupture, releasing a cloud of gas and oil. When the cloud reached an ignition source, the fire flashed back to the release and caused two fatalities and serious burn injuries to several others. The fire could not be extinguished because the source of the fire could not be isolated, as P&IDs were not up to date and the plant tried to keep essential gas supplies flowing.

The Tradeoff. Among the many causes and contributing factors cited in the official report was a decision to relocate experienced technical staff from inside the plant to the company headquarters and to reduce the number of plant supervisors and operators. As one paper put it, "The physical isolation of engineers from the plant deprived operations personnel of engineering expertise and knowledge which previously they gained through interaction and involvement with engineers on site. Moreover, the engineers themselves no longer gained an intimate knowledge of plant activities." (Kenney, Boult, and Pitblado, 2000) This put more responsibility on the shoulders of the remaining operators. When making the decision Esso did not adequately evaluate the tradeoffs involved in the staff restructuring. A management of organizational change review, a form of hazard identification and risk analysis, should have been done to address the impact of the organizational change.

Guidelines for Managing Process Safety Risks During Organizational Change (CCPS 2013) and *Guidelines for Defining Process Safety Competency Requirements* (CCPS 2015a) provide a methodology and a tool, respectively, for analyzing organizational changes.

12.2.4 Make the Decision - Failure to understand uncertainty

Fukushima Nuclear Plant, Explosion, Japan, 2011. This incident was described in Section 6.4 – *Overconfidence*. A broader look at the earthquake record would have shown that the original design basis did not include the possible uncertainty in the estimate. As noted in Section 6.4, the inadequate design basis was recognized by the power company and steps were being taken to improve the design. They were not in time, obviously. One could argue that failure to understand uncertainty is a cause of overconfidence.

12.2.5 Make the Decision – Failure to do risk identification and Failure to probe risk tolerance

ConAgra (2009) and Kleen Energy (2010), natural gas explosions. Natural gas explosions occurred at ConAgra Foods in North Carolina and Kleen Energy Systems in Connecticut within eight months of each other, killing 10 people and injuring over 100 others. The explosion at ConAgra also caused the release of about 18,000 pounds of ammonia to the surrounding environment. Both incidents caused extensive physical damage to buildings. Both were caused by the practice of purging air previously used for pressure testing of new natural gas lines with natural gas. The purge of natural gas was vented into confined areas with no monitoring, no control of ignition sources, and no control of access to minimize people exposed to the hazard. During their investigation of these incidents, the CSB found at least four other similar incidents of this nature (CSB 2009c and CSB 2010).

Although purging of new gas lines with natural gas was an accepted practice at many companies, about half of companies surveyed by the CSB used safer methods, such as pigging with air or nitrogen, nitrogen purges, steam blows, and chemical cleaning. If the companies using natural gas purges had identified the hazard of natural gas purging and/or evaluated the risks vs. their corporate risk tolerance criteria (if they had one), they likely would have chosen one of the safer methods, or at least implemented additional safeguards.

12.2.6 Make the Decision - Failure to recognize linked decisions

Facility siting. Facility siting is an example of linked decisions. Locating a new process unit, storage, or a building in a plant can affect other units

BP Texas City explosion, 2005. This incident was briefly described in Section 2.4.1. To summarize, during startup of an isomerization unit, a raffinate splitter column overflowed. The overflow was directed to a blowdown drum and stack, resulting in a large release of hydrocarbons into the nearby area. The hydrocarbons found an ignition source, and the resulting explosion killed fifteen people, injured over 170, and caused major damage to the isomerization unit and nearby units.

There were a series of decisions that contributed to the fifteen fatalities from this explosion. The first was not to replace the blowdown drum and stack. The blowdown drum was installed in 1950. By the 1970's use of blowdown stacks open to the atmosphere was no longer considered good practice. By 1986, the refinery owners' internal standards called for blowdown stacks be replaced by closed systems when major modifications were made to the unit. Modifications were made in 1995 and 2002, and each time a decision was made not to replace the blowdown drum. This decision let a hazard to units near the blowdown drum continue to exist. (If this choice was not a deliberate one it illustrates the point that not making a decision is a decision in itself. One could call it a non-decision decision.)

The second decision, or missed opportunity, was to not relocate light wood trailers that were in an area where all the fatalities occurred. This decision was made even though an internal standard required such trailers to be at least 350 feet from a process unit unless an MOC review was conducted to approve the location. The MOC review was inadequate in this case, leading to a decision to leave the trailers in place.

12.3 LESSONS AND SUMMARY

Making good process safety risk decisions is more and more necessary in an age of ever-increasing societal expectations as to the safe operations of processing facilities. Major accidents continue to occur, and this means that acute risk decision making is not yet good enough and needs greater attention and more thorough analysis. This chapter reviewed a series of incidents that could have been avoided if the risk decision process had been followed. In some cases, the analysis of the decision process may have seemed to be an example of "second guessing" what the companies involved did. However, an inevitable outcome of any major incident will be the subject of second guessing by regulators and the public. The impact is magnified by 24/7 news coverage. Consider this list of incidents that have occurred recently that received a great deal of coverage:

- 2006 Buncefield explosion, GBR. Overflow of a storage tank resulted in a large explosion and fire that destroyed 20 storage tanks, a nearby car park, and burned for five days. A large increase in the throughput was not recognized as a change and reviewed to see if the control systems and staffing were adequate for the increased throughput. This is an example of a non-decision decision.
- 2010 - Macondo Well explosion and environmental release, Gulf of Mexico. A decision was made to proceed with closure of the well even though pressure tests of the cement plug showed it was not working. Rig operators repeated different tests until they got a result that they thought showed is it was okay to proceed,
- 2011 Chevron refinery fire Richmond, CA. Rather than replace a line with an alloy of steel that could resist sulfidation corrosion, the plant decided to rely on corrosion monitoring for wall thinning. A 4-foot section of line was overlooked when installing the monitoring stations, and that section failed as personnel were trying to stop a leak. The hydrocarbon release resulted in a large fire, with the plumes visible for miles. The California EPA tightened refinery regulations as a result of this incident.
- 2013 West, Texas fertilizer plant explosion. The local community grew and became close to the storage site of large amounts of ammonium nitrate (another example of a non-decision decision).
- 2014 Freedom Industries, West Virginia. Environmental release into the Elk River that contaminated drinking water for weeks.
- 2015 Tianjin, China explosion. An ammonium nitrate explosion killed 170 people, injured about 800, and left a 100 m diameter crater. The operator, RHIL, decided to use political influence to circumvent existing rules and permit systems (123 people were arrested after this incident) (Trembley 2016, Hernandez 2016, and Huang & Zhang 2015).

These incidents all received intense public scrutiny and regulatory intervention. These incidents also damaged the company's reputation, as well as the chemical and petrochemical industry's reputation.

The above list seems to paint a disappointing picture of the CPI's ability to make good decisions. One must remember, however, that it is almost always the high consequence events that make the news. To learn of successes achieved by the application of risk decision tools, one can look through the examples presented at symposiums where companies share both their successes and failures (in the form of case histories). In this book several papers were mentioned that provided examples of successful application of risk decision tools:

- Alspach, J. and Biancji,R.J. 1984, Safe handling of phosgene in chemical processing, Plant/Operation Progress (now Process Safety Progress), Vol. 3 No. 1, January, pp. 40-42.

- Chastain, W., Delanoy, P., Devlin, C., Meuller, T., Study, K., 2017, Beyond HAZOP and LOPA: Four different company approaches, Process Safety Progress, Vol. 36, No. 1, p. 38-53, March.

- Davis, Lloyd and Ness, Albert, 1999, Using quantitative risk assessment to develop a cost-effective spill prevention program, Process Safety Progress, V. 18, No. 4, pp 211-213, Winter.

- Garland, R. Wayne, 2010, Quantitative Risk Assessment Case Study for Organic Acid Processes, Process Safety Progress, Vol. 29, No. 3, p. 247-253, September.

- Gowland, R. 1996, Applying inherently safer concepts to a phosgene plant acquisition, Process Safety Progress, Vol. 15 No. 1, Spring, pp. 52-57.

- Hendershot, D.C., Sussman, J.A., Winkler, G.E., Dill, G.L., 2006, Implementing inherently safer design in an existing plant, Process Safety Progress, Vol. 25, No. 1, p. 52-57, Marc.

- Myers, P., Morgan, R & Flamberg, S. 1997, Toxic Hazard Reduction with Passive Mitigation Systems Process Safety Progress, Vol. 15 No. 1, Spring, pp. 225-230.

Literature searches can reveal many more examples.

Following a rigorous decision process, applying the correct decision analysis and decision-making tools, and considering tradeoffs, linkage to other decisions, and understanding the uncertainty in the decision process will lead to improved outcomes.

REFERENCES

Alspach, J. and Biancji,R.J. 1984, Safe handling of phosgene in chemical processing, *Plant/Operation Progress (now Process Safety Progress)*, Vol. 3 No. 1, January, pp. 40-42.

Baker Panel, 2007, *The Report of the BP refineries Independent Safety review* panel, http://www.csb.gov/assets/1/19/Baker_panel_report1.pdf

Baybutt 2007, An improved risk graph approach for determination of Safety Integrity Levels (SILs), *Process Safety Progress*, Vol. 26, No. 1, pp 66-762, March.

Baybutt 2014, The use of risk matrices and risk graphs for SIL determination, *Process Safety Progress*, Vol. 33, No. 2, pp 179-182, June.

Bridges and Dowell 2016, 'Identify SIF and specify necessary SIL, and other IPLs, as part of PHA/HAZOP', *Proceedings of the 12th global congress on process safety*, Center for Chemical Process Safety, New York.

Birtwistle, John 2001, *Hickson Pharmachem Ltd, Runaway Reaction Explosion and Fire*, American Institute of Chemical Engineers, Center for Chemical Process Safety, New York, NY.

CAIB 2003, *Report of Columbia accident investigation board, volume I*, Columbia Accident Investigation Board, Government Printing Office, Washington, D.C. (http://www.nasa.gov/columbia/home/CAIB_Vol1.html).

CCPS 1989, *Guidelines for process equipment reliability data, with data tables*, American Institute of Chemical Engineers, Center for Chemical Process Safety, New York, NY.

CCPS 1995, *Tools for making acute risk decisions with chemical process safety applications*, American Institute of Chemical Engineers, Center for Chemical Process Safety, New York, NY.

CCPS 1995a, *Guidelines for Process Safety Documentation*, American Institute of Chemical Engineers, Center for Chemical Process Safety, New York, NY.

CCPS 1996, *Inherently Safer Chemical Process: A Life Cycle Approach*, American Institute of Chemical Engineers, Center for Chemical Process Safety, New York, N

CCPS 1999, *Guidelines for chemical process quantitative risk analysis*, 2nd Ed., American Institute of Chemical Engineers, Center for Chemical Process Safety, New York, NY.

CCPS 1999a. *Guidelines for process safety in batch reaction systems*, American Institute of Chemical Engineers, Center for Chemical Process Safety, New York, NY.

CCPS 2000, *Guidelines for chemical process quantitative risk analysis, 2nd edition*, American Institute of Chemical Engineers, Center for Chemical Process Safety, New York, NY.

CCPS 2001, *Layer of protection analysis: simplified process risk assessment*, American Institute of Chemical Engineers, Center for Chemical Process Safety, New York, NY.

CCPS 2001a, *Revalidating process hazard analyses*, American Institute of Chemical Engineers, Center for Chemical Process Safety, New York, NY.

CCPS 2003, *Essential practices for managing chemical reactivity hazards*, American Institute of Chemical Engineers, Center for Chemical Process Safety, New York, NY.

CCPS 2007. *Guidelines for risk based process safety*, American Institute of Chemical Engineers, Center for Chemical Process Safety, New York, NY.

CCPS 2007b. *Guidelines for safety and reliable instrumented protection systems*, American Institute of Chemical Engineers, Center for Chemical Process Safety, New York, NY

CCPS 2008, *Guidelines for hazard evaluation procedures, 3rd Ed.*, American Institute of Chemical Engineers, Center for Chemical Process Safety, New York, NY.

CCPS 2008a, *Incidents that define process safety*, Center for Chemical Process Safety of the American Institute of Chemical Engineers, New York, NY.

CCPS 2008b, *Guidelines for management of change for process safety*, Center for Chemical Process Safety of the American Institute of Chemical Engineers, New York, NY.

CCPS 2009, *Inherently safer chemical processes: a life cycle approach*, American Institute of Chemical Engineers, Center for Chemical Process Safety, New York, NY.

CCPS 2009a, *Guidelines for developing quantitative safety risk criteria*, American Institute of Chemical Engineers, Center for Chemical Process Safety, New York, NY.

CCPS, 2010, *Guidelines for vapor cloud explosion, pressure vessel bursts, BLEVE and flash fire hazards*, American Institute of Chemical Engineers, Center for Chemical Process Safety, New York, NY.

CCSP 2012. *Guidelines for engineering design for process safety, 2nd edition*, American Institute of Chemical Engineers, Center for Chemical Process Safety, New York, NY.

CCSP 2013. *Guidelines for managing process safety risks during organizational change, 2nd edition*, American Institute of Chemical Engineers, Center for Chemical Process Safety, New York, NY.

CCPS 2014, *Guidelines for enabling conditions and conditional modifiers in layer of protection analysis*, American Institute of Chemical Engineers, Center for Chemical Process Safety, New York, NY.

CCPS 2015, *Guidelines for initiating events and independent protection layers*, American Institute of Chemical Engineers, Center for Chemical Process Safety, New York, NY.

CCPS 2015a, *Guidelines for defining process safety competency requirements*, American Institute of Chemical Engineers, Center for Chemical Process Safety, New York, NY.

CCPS 2018, *Bow Ties in risk ranagement: A concept book for process safety*, American Institute of Chemical Engineers, Center for Chemical Process Safety, New York, NY.

CCPS 2018a, *Recognizing and responding to normalization of deviance*, American Institute of Chemical Engineers, Center for Chemical Process Safety, New York, NY.

Chastain, W., Delanoy, P., Devlin, C., Meuller, T., Study, K., 2017, Beyond HAZOP and LOPA: Four different company approaches, *Process Safety Progress*, Vol. 36, No. 1, p. 38-53, March.

Considine, M. and Hall, S.M., 2009, The major accident risk (MAR) process – developing the profile of major accident risk for a large multi national oil company, *Process Safety and Environmental Protection*, Vol. 87, No. 1, p. 59-63, January.

CSB 2000. *Chemical manufacturing incident - April 1998*, U.S. Chemical Safety and Hazard Investigation Board, Investigation Report, Report No. 1998-06-I-NJ, August. (http://www.csb.gov/investigations).

CSB 2002, *Improving reactive hazard management*, US Chemical Safety and Hazard Investigation Board Investigation Report, No. 2001-01-H, October 2002.

CSB 2002a, *Explosion and fire, First Chemical corporation*, U.S. Chemical Safety and Hazard Investigation Board, Investigation Report, Report No. 2003-01-I-TX, October 13, (http://www.csb.gov/investigations).

CSB 2003, *Hazards of nitrogen asphyxiation*, U.S. Chemical Safety and Hazard Investigation Board, Safety Bulletin, No. 2003-10-B, June, (https://www.csb.gov/hazards-of-nitrogen-asphyxiation/)

CSB 2006, *Toxic chemical vapor cloud release*, U.S. Chemical Safety and Hazard Investigation Board, Investigation Report, Report No. 2004-09-I-GA, April, (http://www.csb.gov/investigations).

CSB 2007. *Refinery explosion and fire. BP Texas City, Texas*, U.S. Chemical Safety and Hazard Investigation Board, Investigation Report, Report No. 2005-04-I-MS, March 23, (http://www.csb.gov/investigations).

CSB, 2009, *T2 Laboratories, Inc. runaway reaction*, U.S. Chemical Safety and Hazard Investigation Board, Report No. 2008-3-I-FL, September.

CSB 2014. *Chevron Richmond refinery pipe rupture and fire*, Chevron Richmond Refinery #4 Unit, Richmond, CA, U.S. Chemical Safety and Hazard Investigation Board, Investigation Report, Report No. 2012-03-I-CA, October. (http://www.csb.gov/investigations).Gawande, A. 2009, *The Checklist manifesto*, Henry Holt and Company, New York, NY, 2009.

CSB 2014a, *Explosion and fire at the Macondo Well; Vol 3, Human, organizational and safety system factors of the Macondo blowout*, Chemical Safety and Hazard Investigation Board, Case Study, Report No. 2010-10-I-OS, June 5. (http://www.csb.gov/investigations).

Davis, Lloyd and Ness, Albert, 1999, Using quantitative risk assessment to develop a cost effective spill prevention program, *Process Safety Progress*, V. 18, No. 4, pp 211-213, Winter.

DIN 1994, (German Institute for Standardization) V 19250, *Control technology: Fundamental safety aspects to be considered for measurement and control equipment*, DIN, Berlin, Germany.

Dobelli, Rolf, 201, *The Art of thinking clearly*, Harper Collins, New York, NY.

DOD 2000, *Standard practice for system safety*, MIL-STD-882D, U.S, Department of Defense, February. (http://www.safetycenter.navy.mil/instructions/osh/milstd882d.pdf)

Gawande 2009, The Checklist Manifesto, Ault Gawande, Henry Holt and Company, New York, NY, 2009.

Garland, R. Wayne, 2010, Quantitative Risk Assessment Case Study for Organic Acid Processes, Process Safety Progress, Vol. 29, No. 3, p. 247-253, September.

Govindarajan and Terwilliger 2012, Yes, you can brainstorm without groupthink, *Harvard Business Review, Digital Article*, July 25, (https://hbr.org/2012/07/yes-you-can-brainstorm-without).

Gowland, R. 1996, Applying inherently safer concepts to a phosgene plant acquisition, *Process Safety Progress*, Vol. 15 No. 1, Spring,pp. 52-57.

Hammond, J.S., Keeney R.L. & Raiffa H. 1999, *Smart choices: a practical guide to making better life decisions*, Harvard Business School Press, Boston, 1999.

Hansen 2103, How John F. Kennedy changed decision making for us all, *Harvard Business Review, Digital Article*, November 22, (https://hbr.org/2013/11/how-john-f-kennedy-changed-decision-making).

Hendershot, D.C., 1996, Risk guidelines as a risk management tool, *Process Safety Progress*, Vol. 15, No. 4, p. 213-218, Winter.

Hendershot, D.C., Sussman, J.A., Winkler, G.E., Dill, G.L., 2006, Implementing inherently safer design in an existing plant, *Process Safety Progress*, Vol. 25, No. 1, p. 52-57, March.

Hernandez, J.C. 2016, Tianjin explosions were result of mismanagement, China finds, New Yok Times, Feb. 5.

Hopkins, Andrew, *Failure to learn: the BP Texas City refinery disaster*, CCH Australia, Sidney, 2008.

HSE (1998), *Development of methods to assess the significance of dominnoffects from major hazard sites*, Health and Safety Executive (UK), ISBN 0 7176 2151 0, Her Majesty's Stationary Office, Norwich, UK, 2001. http://www.hse.gov.uk/research/crr_pdf/1998/crr98183.pdf

HSE (2001), *Reducing risk, protecting people: HSE's decision-making process*, Health and Safety Executive (UK), ISBN 0 7176 2151 0, Her Majesty's Stationary Office, Norwich, UK, 2001. http://www.hse.gov.uk/risk/theory/r2p2.pdf

Huang, P. and & Zhang, J. 2015, Facts related to August 12, 2015 explosion accident in Tianjin, China, Process Safety Progress, Vol.34, No.4, December.

IAEA 2015, *The Fukushima Daiichi accident*, International Atomic Energy Agency, Vienna, Austria 2015.

IChemE 2018, *The wisdom of Trevor Kletz – the 'founding father' of inherent safety*, (https://ichemeblog.org/2015/05/15/the-wisdom-oftrevor-kletz-the-founding-father-of-inherent-safety-day-353/)

IEEE 1984. Std. 500-1984: *IEEE guide to the collection and presentation of data for nuclear power generating stations*, Institute of Electrical and Electronics Engineers, New York, NY, 1984.

IEEE 2007. Std. 493-2007: *Recommended practice for the design of reliable industrial and commercial power systems*, Institute of Electrical and Electronics Engineers, New York, NY, June 2007.

ISO 2009, *ISO 31000, Risk management – principles and guidelines*, International Standards Organization, Geneva, 2009.

G. D. Kenney, M. Boult, R. M. Pitblado 2000, *Lessons for Seveso II from Longford Australia*, EU Safety Conference: Implementation of the Seveso II Directive, Athens, Greece.

Kepner, C.H., and Tregoe, B.B., *The New rational manager*, John Martin Publishing Ltd., London, 1981.

Kytomaa et al. 2019, An integrated method for quantifying and managing extreme weather risks and liabilities for industrial infrastructure and operations, *Process Safety Progress*, Vol. 38 No. 2, Spring.

Lees, F.P. 1996, *Loss prevention in the process industries – hazard identification, assessment and control'*, Volume 3, Appendix 4, Butterworth-Heinemann, Elsevier, Oxford, UK.

Lees (2004), Lees' *Loss prevention in the process industries, 4th ed.*, Mannan, S. (editor), Butterworth-Heinemann, Elsevier, Oxford, UK.

Lewis, Michael 2017, *The Undoing Project, a friendship that changed our minds*, W.W. Norton & Company, New York, N.Y.

Markman 2015, The Problem-Solving process that prevents groupthink, *Harvard Business Review, Digital Article,* November 25, 2015, (https://hbr.org/2015/11/the-problem-solving-process-that-prevents-groupthink).

Mouawad, Jan, 2010, New culture of caution at Exxon after Valdez, *New York Times*, July 12.

Murphy, J. and Connor, J. 2012, Beware of the black swan: the limitations of risk analysis for predicting the extreme impact of rare process safety incidents, *Process Safety Progress*, Vol.31, No.4, p. 330-333

Myers, P., Morgan, R & Flamberg, S. 1997, Toxic Hazard Reduction with Passive Mitigation Systems *Process Safety Progress*, Vol. 15 No. 1, Spring, pp. 225-230.

OSHA 1992, *Process safety management of highly hazardous chemicals (29 CFR 1910.119);* Federal Register 1992, Vol. 57, No. 36, February 24. https://www.osha.gov/pls/oshaweb/owadisp.show_document?p_table =STANDARDS&p_id=9760

Pitblado, R. , Bardy, M., Nalpanis, P., Crosshwaite, P., Molazemi, K., Bekjaert, M., Raghunathan, V., 2012, International comparison on the application of societal risk criteria. *Process Safety Progress*, Vol. 31, No. 4, p. 363-368, Oct. 5.

Pitbaldo, R. and Weijand, P. (2014), Barrier diagram (Bow Tie) quality issues for operating managers, *Process Safety Progress*, Vol.33, No.4, p. 355-361, December.

Rosen 2015, Rosen, Robert, The Ten commandments of risk based process safety, *Process Safety Progress*, Vol. 34, No. 3, p. 212-213, September.

Sanders, R. E. 1993, Don't become another victim of vacuum," Chemical Engineering Progress, Vol 89 (9), pp. 54-57, .

Taleb, N.N., 2010, The Black swan: the impact of the highly improbable, Random House Digital, New York.

TNO 2007. Purple Book - CPR 18E - *Guidelines for quantitative risk assessment*, TNO 2005, Netherlands Organization for Applied Scientific Research, The Hague, Netherlands, 2005.

Topalovic, Peter, Krantzenberg, Gail, 2013, *Responsible care: A case study*, De Gruyter, Berlin, GE.

Tremblay, J. 2016, Chinese investigators identify cause Of Tianjin explosion, Chemical and Engineering News, February 8.

UK 1975, *The Flixborough Disaster – Report of the Court of Inquiry*, Her Majesties Stationary Office, London, UK.

https://www.icheme.org/communities/special-interest-groups/safety%20and%20loss%20prevention/resources/~/media/Documents/Subject%20Groups/Safety_Loss_Prevention/HSE%20Accident%20Reports/The%20Flixborough%20Disaster%20-%20Report%20of%20the%20Court%20of%20Inquiry.pdf

INDEX